I0463835

BIOLOGY BYTES

Digestible Essays on
Stem Cells and
Modern Medicine

Dr. Teisha J. Rowland

Also by DR. TEISHA J. ROWLAND

BIOLOGY BYTES: Digestible Essays on Animals Both Commonplace and Bizarre

Biology Bytes

http://www.biology-bytes.com

All new essays on varied biology topics

All Things Stem Cell

http://www.allthingsstemcell.com

Continuing commentaries on stem cell research

© 2009 - 2013 by Teisha J. Rowland.

All rights reserved.

Printed in the United States of America.

Essays originally appeared in the Santa Barbara Independent, Inc., in 2009 – 2011.

First published as a paperback in 2013.

No part of this book may be reproduced or transmitted in any form or by any means (electronic, mechanical, or any other information retrieval or storage system), except for the purpose of reviews, without permission from the author. The mention of or reference to any companies, individuals, or products in these pages is not a challenge to the trademarks or copyrights concerned.

Softcover ISBN: 978-1-304-27574-5

Editor: Dr. Andrew J. Bonham

To order additional copies of this book, order online at
http://www.biology-bytes.com

Contents

Chapter 4: Better Understanding Cancer

Chapter 5: What You Eat

Preface

A few years ago, I had an idea for a biology column in the local newspaper. All sorts of biology-related topics would be covered, from the history of pets to the latest stem cell-based medical breakthroughs and much more in between. The column would explore the biology that's all around everyone, all of the time. (But that often gets overlooked, or misunderstood.) It would strive to make even the most complex biological topics, like bioengineering organs or making vaccines against cancer, become something understandable. And, maybe if I did it right, it'd ignite interest and inquiry. This is how, in 2009, I pitched the idea of a column on varied biology topics to the *Santa Barbara Independent*. The *Independent* accepted my pitch and "Biology Bytes" was born. In October 2009, it started exploring life one digestible tidbit at a time.

Biology has always been my passion. When I started writing "Biology Bytes," I was a Ph.D. graduate student at the University of California, Santa Barbara (UCSB), in the department of Molecular, Cellular, and Developmental Biology (MCDB). There I studied stem cells. More specifically, in the laboratory of Prof. Dennis O. Clegg, I investigated how to persuade different types of stem cells to turn into specialized retinal cells that could be used to treat the disease of macular degeneration. I was excited about stem cells and their potential. In fact, I was so enthused that after working with them all day, I'd then go home to learn more about stem cells, often types completely unrelated to my thesis project. The more I discovered, the more I wanted to share my interest in these amazing cells with other people. This is why, earlier in 2009, I had created a blog about them, "All Things Stem Cell" (http://www.AllThingsStemCell.com). Like "Biology Bytes," in my blog I tried to make complex scientific topics as accessible as possible. Later I adapted several blog posts to be articles for "Biology Bytes."

Throughout the rest of graduate school, I regularly wrote one article a week for "Biology Bytes." Evenings and weekends, when I wasn't doing my stem cell research in the lab, I would often be reading and writing. Stacks of books on all kinds of biology topics would clutter my desk. I ended up writing 67 articles total. My last article was published in May 2011, two months before I received my Ph.D.

from UCSB and moved back to Colorado, my home state. It was difficult to end the series. I had taken great pleasure in writing those articles (as exhausting as it had been at times due to being a full-time graduate student), and I knew the column had caught the interest of many readers who enjoyed reading a bit of real science.

After mulling it over, I decided to publish the articles as a book. But I didn't want to publish the articles as they were – since some articles were nearly two years old by this point, I wanted to make sure the information within them was still current. Consequently I reviewed each article, updated its content, and included any additional relevant information on the topic. It was challenging at times because breakthroughs on a given topic seemed to pop up in the news all of the time, and I had to include any that would be of interest. As I went through this process, I found that it made more sense to publish the essays in two separate books, one containing the more medically-oriented essays and another with the animal-related ones. This is how *Biology Bytes: Digestible Essays on Stem Cells and Modern Medicine* and *Biology Bytes: Digestible Essays on Animals Both Commonplace and Bizarre* both came about, respectively. So while the essays in this book were originally published in the *Independent,* here they contain a significant amount of new content that was not in the original articles.

The essays in this book have been organized according to topic, with each chapter being a different topic. The book may be read straight from the first chapter to the last, or one topic at a time. The chapters may even be read as standalone essays, but when read together in a chapter they certainly work to build a foundation of knowledge on the topic. In case the reader skips around, the chapters cross-reference each other, pointing to where background gaps may be filled.

To read more fascinating and digestible biology-based tidbits, visit the ongoing exploration at the Biology Bytes website at http://www.biology-bytes.com.

Acknowledgements

I would like to start by thanking the *Santa Barbara Independent* for letting my idea of a biology column become reality. Thanks to Matt Kettmann for giving me a chance and being supportive of this endeavor. Thanks to Martha Sadler for editing my articles and, in doing so, helping me figure out what a science article in a newspaper should be and what it shouldn't be.

Thanks to the University of California, Santa Barbara, for having an atmosphere that encourages inquiry, as well as a great library. Thanks to my Ph.D. advisor, Prof. Dennis O. Clegg, who shared with me his interest in stem cells and, more importantly, a boundless enthusiasm for, and curiosity about, cool science in general. Thanks to Prof. Kathy Foltz, whose approach to science education inspired my own, and support of my science writing pursuits showed me that my non-traditional dreams are attainable.

Thanks to scientists who got my feet wet before graduate school and helped instill in me the importance of improving science education, especially Dr. Eduardo Orias and Dr. William B. Wood. Thanks to my third grade teacher, Marian Shellaberger, who got me excited about biology. Thanks to Science Buddies, an amazing company that hired me after receiving my Ph.D. so that I can work with other scientists to improve the science experience for K-12 students, all while having fun learning about, and experimenting on, different science topics. Special thanks to all science educators – thank you for working hard to ignite that interest and inquiry that's vital to our survival.

Thanks to my loving parents, Bob and Meg Rowland. Thank you for both being supportive, encouraging, and proud of me. Thanks also to my sister, Erin, who explored fields, ponds, and raising goats, chickens, guinea pigs, and so many other pets with me.

Most of all, thanks to my loving husband, Prof. Andrew James Bonham. Thank you for undertaking many journeys with me so far, and many more to come. Thank you for helping being so many things for me: scientific collaborator, remarkable and patient editor, supportive best friend and companion, entertainer, and my love. Thank you for helping me improve myself, and for sharing your life with me.

CHAPTER 1

STEM CELLS

The Stem Cell Family

A bigger-picture view

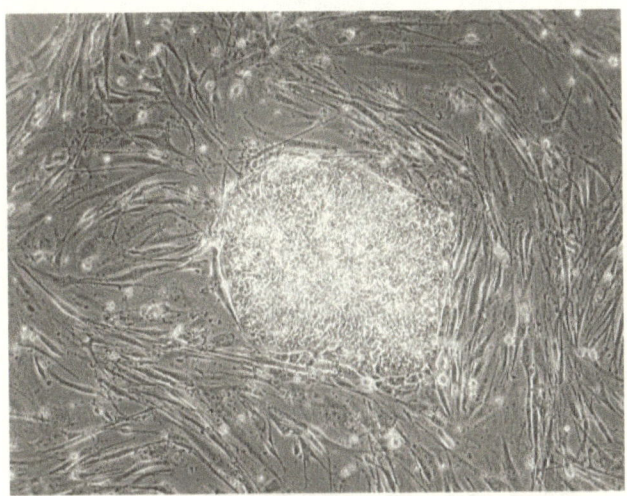

Human embryonic stem cells, shown growing in a "colony" in the middle of the picture above, have been an essential addition to the ever-growing stem cell family. They are also one type of stem cell that I used extensively in my research as a graduate student.

With all the breaking news stories that come out on cutting-edge stem cell findings each week, it can be easy to lose sight of the bigger picture. Yes, the stem cell family, which includes all of the varieties of stem cells we've discovered so far, is very large, and growing larger with new children, cousins, uncles, and aunts being discovered or created all the time. But a key feature they all share is their potential to improve our lives.

Our understanding of these cells and their incredible potential for treating diseases, fighting cancers, healing wounds, and, in essence, saving lives, has grown hugely since we first unknowingly used them in World War II. However, the more we learn about them the more we realize we have yet to understand. The following sections in this chapter on stem cells strive to explore the different stem cell types in detail, including their biology, history, potential, clinical applications, and numerous remaining questions. However, the ways in which the different types of stem cells came to be accepted into the stem cell family is itself an interesting story, and one that can help paint a useful bigger picture. This story will provide context for the many members of the stem cell family that will be discussed throughout the rest of this chapter.

HEMATOPOIETIC STEM CELLS

The first time stem cells were successfully used to treat patients was during World War II. However, at the time, people did not know they were using stem cells. During World War II, people exposed to lethal doses of radiation were given bone marrow transplants that, somehow, could cure them. Much later it was discovered that the responsible agents in the bone marrow were hematopoietic stem cells (HSCs), which will be discussed in detail later in this chapter.

Because hematopoietic stem cells are rapidly growing cells, they're particularly damaged by exposure to radiation. Consequently, a radiation victim may need a transplant of healthy HSCs to replace their own. As this history makes apparent, HSCs reside in bone marrow (as well as other tissues), making bone marrow a good HSC transplantation source.

But what do these HSCs do exactly? They're quite important. They can turn into all the different types of blood cells in the body. This is why transplants of HSCs (from bone marrow) are also used to treat cancers of the hematopoietic (blood) system, such as leukemia or lymphoma. They've even been successfully used in treating a patient with both leukemia and HIV. Today hematopoietic stem cells are one of the few adult stem cells that are widely used clinically.

EMBRYONIC STEM CELLS

While bone marrow donor centers were being established in the 1980s, another stem cell family tree branch was developing that would draw much attention: Nearly 30 years ago, embryonic stem cells were isolated from early-stage mouse embryos. It was not until 1998 that the same feat was accomplished with human embryos by James Thomson, who holds faculty appointments at the University of Wisconsin and the University of California at Santa Barbara (UCSB). From 2006 to 2011, I was a graduate student at UCSB in the laboratory of Dennis Clegg, who helped recruit Thomson to UCSB and vitalized stem cell research and collaborations on campus through the creation of the Center for Biology and Engineering at UCSB.

These human embryonic stem cells (hESCs), which are discussed in detail later in this chapter, are isolated from early stage embryos. Technically called blastocysts, these are embryos that have not yet implanted in the uterus. They have existed for only five days after fertilization and contain only about 150 cells. One of the most promising qualities of hESCs is their ability to become virtually any cell type; this potential means they are "pluripotent." Although it was not until 2009 that these newer family members received F.D.A approval for use in a clinical trial, they have since been approved in other clinical trials.

MESENCHYMAL STEM CELLS

Around the same time that researchers were figuring out how to isolate embryonic stem cells from humans, yet another large group of stem cells was finally admitted into the stem cell family. Although researchers reported the existence of mesenchymal stem cells (MSCs) as early as the 1960s and 1970s, which will be discussed in detail later in this chapter, they were excluded from the stem cell family for decades. Why the prejudice? MSCs had somewhat questionable origins; they're most commonly harvested from adipose tissue (fat) or bone marrow. Researchers already knew bone marrow was home to hematopoietic stem cells and it was difficult to accept that they shared their home with yet another group of stem cells. But by the late 1990s, MSCs were firmly established and allowed into the family.

MSCs hold great potential for the field of regenerative medicine, as they can become many different types of cells (they're "multipotent"), most typically bone, cartilage, and fat cells. In 2000, a new member was added to the mesenchymal stem cell branch, when researchers discovered that teeth are home to dental pulp stem cells.

NEWEST FAMILY ADDITIONS

During the last decade, a number of new stem cell types have joined the every-growing family, through discovery or creation. In 2007, stem cells were found to reside in menstrual blood (more on that later). It's unclear exactly what branch these "endometrial regenerative cells" belong to in the stem cell family; they have similarities with both embryonic stem cells and mesenchymal stem cells. Nonetheless, because they can be obtained in a non-invasive manner, and in large quantities, they offer much potential for cellular therapies.

2007 also saw one of the most game-changing developments in the stem cell field; researchers learned how to create cells *like* embryonic stem cells, but instead of coming from an embryo these cells can be created from adult cells, potentially cells from any tissue in the human body. These cells, called induced pluripotent stem cells (iPSCs), which we will focus on later, are created by forcing adult cells to produce proteins that are specific to embryonic stem cell functions, causing the adult cells to look and act like ESCs. iPSCs have significantly altered the field ethically, as they have the utility of embryonic stem cells but do not require harvesting cells from a blastocyst. They also alter it clinically: It's now possible, in theory, to grow patient-specific pluripotent cells.

But the idea of changing a mature cell's identity had already been around for many decades. Since the 1930s, researchers have been exploring and developing a technology that would later be called somatic cell nuclear transfer (SCNT). In many ways, SCNT technology formed the basis for the idea of iPSCs. SCNT has been used to clone animals, such as the famous Dolly the sheep in 1997, and may be used for therapeutic cloning (i.e., cloning a patient's tissues or organs to use in a transplant).

Like SCNT, a technique called "direct reprogramming" has also been used for decades to change a mature cell's fate. Since the late 1980s, researchers had used direct reprogramming to, for example, make different types of adult cells turn into muscle cells. They did this by making the adult cells produce proteins essential to the identity of muscle cells (we will examine direct reprogramming in detail further on). It was found that other cells could have their "established" identities altered in similar ways, with a group in 2008 reporting that some pancreas cells (exocrine cells) could be turned into other, insulin-producing pancreas cells (beta-cells), which may hold promise for treating diabetes. In 2010, another group found that fibroblast cells (cells active in connective tissue) taken from mouse tails could be made into nerve cells, or neurons.

In 2011, progenitor cells, which are like a limited stem cell, received closer examination by the stem cell community. Specifically, it was found that in male pattern baldness it may be progenitor cells, and not stem cells, that are responsible for causing the condition. Because of this finding, progenitor cells are likely to receive greater inclusion in future family discussion. These cells and this discovery will be discussed more later in this chapter.

SO WHAT HAVE STEM CELLS DONE FOR YOU LATELY?

Although many stem cell therapies are still in their infancy, in the last couple years there have been several important publications on the successful use of novel stem cell treatments in patients.

ENGINEERING ORGANS

In 2008, Claudia Castillo had her near-collapsed bronchus replaced by a trachea from a cadaver that had her own cells grown on it, which will be discussed later in this chapter. These cells included chondrocytes (cartilage cells) that were made from mesenchymal stem cells that had been taken from her bone marrow; turning these MSCs into chondrocytes only took researchers three days. Since 2008, this technique has been improved upon and successfully used in other patients.

TREATING HIV

Although hematopoietic stem cells have been used since World War II to treat victims of radiation, a potential, significant new application was reported in 2009: Hematopoietic stem cells were used to successfully treat a patient with HIV, discussed later in this chapter, although the procedure is currently risky and a great deal of additional research is necessary for it to be widely accepted and used.

TREATING SPINAL CORD INJURIES

2009 also saw the first FDA-approval of the use of human embryonic stem cells in a clinical trial. Hans Keirstead, who will make a featured appearance, developed the therapy to be used in the trial; Keirstead and colleagues made hESCs become oligodendrocytes and then showed that these cells could help cure spinal cord injuries in animals. Unfortunately, the trial has been stalled due primarily to financial concerns, but it undoubtedly helped pave the way for other clinical trials using hESC-based therapies to receive F.D.A-approval in 2010, and surely others to follow.

FIGHTING CANCER

Stem cells have also helped us better understand cancer, through investigation of the emerging idea of the cancer stem cell, which will be discussed in the next section of this chapter, and the many similarities cancer shares with different members of the stem cell family. Such studies may even help researchers develop better cancer vaccines, which will be discussed in chapter four.

As our understanding of this complex and constantly growing family continues to grow, so too should our understanding of how the medical field can best use the different members to improve our lives.

Further Reading

The following resources may be of interest for those wanting to learn more about the stem cell family:

- Teisha J. Rowland's blog "All Things Stem Cell" at http://www.AllThingsStemCell.com
- The National Institutes of Health's webpage "Stem Cell Information: Frequently Asked Questions (FAQs)" at http://stemcells.nih.gov/info/faqs.asp

- Teisha J. Rowland's webpage "Visual Stem Cell Glossary" at http://www.AllThingsStemCell.com/Glossary

Stem Cells and Cancer

Cancer stem cells seem to play a key role in cancer

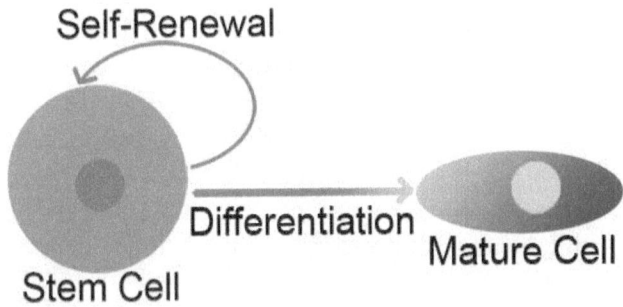

Cancer stem cells and cancer share two key characteristics; both can create more cells, a process called "self-renewal" in stem cells, and both can become different, mature cell types, called "differentiation" in stem cells.

Recently, a link was discovered between two widely-studied medical areas: cancer and stem cells. Cancer stem cells, as their name implies, are stem cells that have been found within cancerous tumors.

When you understand stem cells, it quickly becomes apparent why they might be in tumors. Cancerous tumors are a mixture of many different cell types. If the tumor is originally from one cell gone bad, that single cell must be able to become different types of cells; a key defining characteristic of stem cells is that they can become multiple different cell types, a process called "differentiation."

Another central trait of malignant tumors and stem cells is their proliferative ability; tumors can grow out of control, while stem cells, by definition, can "self-renew," or create more stem cells.

These two shared traits make stem cells a very likely suspect of playing a key role in cancer, and led to the theory of cancer stem cells (CSCs). CSCs have been broadly defined as cells within a tumor that can become other cells (creating the heterogeneity seen in tumors), and self-renew (regenerating the CSC population).

HOW CANCEROUS TUMORS COME TO BE

In order to comprehend how cancer stem cells are created, it is important to understand the theories behind the creation of cancerous tumors. Cancer can occur when mutations have accumulated in certain genes. Genes are turned into protein by our cells and serve a specific function in each and every cell. The genes usually mutated in a cancerous tumor are (1) genes that control cell growth and (2) genes that make sure the cell retains its cell type (i.e. a liver cell does not become a brain cell). There are special names for these genes; they are oncogenes ("onco-" being the prefix for "tumor") and tumor suppressor genes. These genes are often mutated in CSCs.

A mutation in a gene that normally regulates cell growth can cause the cell it belongs to to quickly increase in numbers, creating a large population of mutated cells. This is potentially very dangerous because of how probability works; there is a greater likelihood that a large population of cells will gain a gene mutation than a small population of cells. When enough mutations accumulate in a cell, it can overcome the normal cell growth restrictions so thoroughly that it grows out of control, becoming cancerous.

CANCER STEM CELLS IDENTIFIED

Although the cancer stem cell theory developed in the 1970s, recently it has gained a spotlight in the scientific community. The first functional identification of CSCs was in 1997 in acute myeloid leukemia. Researchers found that although tumors are made up of many different cell populations, there was only one population capable of generating the tumor. This was determined by taking a tumor and breaking it up into its many different cell types and then injecting each separately into mice; one group of cells, the CSCs, could recreate the original tumor, including the shape and specific cell types found in the original. It has been found that CSCs account for only a small percentage of the total number of cells in the tumor, varying from as little as 0.002 percent to around 30 percent, depending on the type of tumor, but it seems to be usually less than 10 percent.

CELL MARKERS

The different cell populations within a tumor can be separated and identified by the proteins they make; cells creating the same sets of proteins can be grouped into one population. Because such proteins are commonly used to identify and categorize cells, they are called cell markers. CSCs from the same type of tumor make the same set of markers. However, CSCs in multiple different cancer types share some of the same cell markers. For example, CSCs found in pancreatic cancer produce a cell marker called CD24. But CD24 is not made by all CSCs; breast cancer CSCs do not produce CD24, but do produce another marker called CD44. Interestingly, CD44 is also made by CSCs in pancreatic cancer and several other cancers: ovarian cancer, head and neck squamous cell carcinoma, and probably others. Another cell marker commonly found in different tumor types is CD133; CSCs in malignant tumors of the brain, prostate, colon, lung, and liver have it.

To complicate matters even further, these cell markers are not found exclusively in cancerous tumors; these markers are often in stem cells in normal, healthy tissue and do not pose any apparent malignant threat. Ordinary stem cells from the brain, intestinal lining, and other tissues, all make the marker CD133. Likewise, CD44 is produced by normal mesenchymal stem cells and bone marrow stem cells. Mesenchymal stem cells (which are in muscle, connective tissue, fat, and fetal liver and lungs) and bone marrow stem cells are both discussed later in this chapter. Overall, it's become clear that some CSCs have the same markers as normal stem cells in the healthy tissue of the tumor origin. What's unclear is what the implications of this are.

However, surprisingly, sometimes CSCs produce cell markers different from the normal stem cells in the healthy tissue they're surrounded by. These CSCs may be migrating from another tissue to the tumor site, or, alternatively, the CSCs may be mutated stem cells that resided in the original, healthy tissue. Because CSCs can become other cell types, they could change which cell markers they produce; CSCs may make a set of markers very different from the cells they were before residing in a cancerous tumor.

As more evidence is reported pointing at the key role of cancer stem cells in the creation of cancerous tumors, it becomes more crucial for researchers to have a thorough understanding of these stem cells. Although the CSC populations can be identified by different cell markers and their ability to create tumors in a mouse model, there is still much about them that is not well understood: how they are created, how their origins are related to non-cancerous stem cells, and whether they are present in all cancerous tumors. As more answers come to light, researchers will be able to answer the most important question: How can we use our knowledge of CSCs to most effectively combat them? Some possibilities are explored in Chapter 4 with the development of cancer vaccines.

Further Reading

The following resources may be of interest for those wanting to learn more about cancer stem cells, cancer, and stem cells in general:

- Teisha J. Rowland's blog "All Things Stem Cell" at http://www.AllThingsStemCell.com
- Teisha J. Rowland's blog article "Cancer Stem Cells: A Possible Path to a Cure" at http://www.AllThingsStemCell.com/2009/07/cancer-stem-cells/
- The National Institutes of Health's webpage "Stem Cell Information: Frequently Asked Questions (FAQs)" at http://stemcells.nih.gov/info/faqs.asp
- Teisha J. Rowland's webpage "Visual Stem Cell Glossary" at http://www.AllThingsStemCell.com/Glossary
- Ning Ning et al.'s article "Cancer Stem Cell Vaccination Confers Significant Antitumor Immunity" at http://dx.doi.org/10.1158/0008-5472.CAN-11-1400

Human Embryonic Stem Cells

Fifteen years of discovery, controversy, and potential

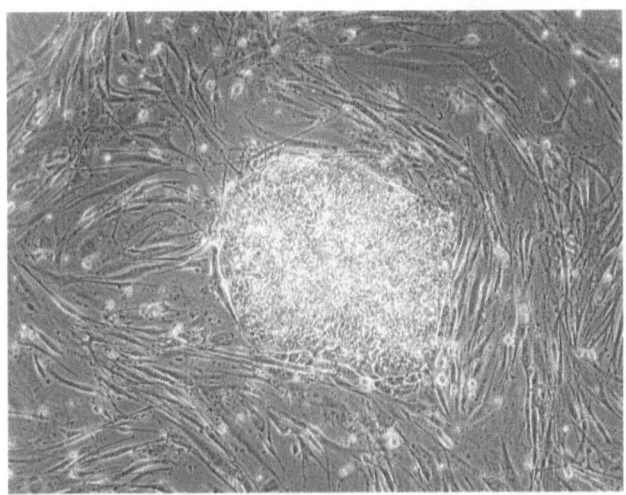

Human embryonic stem cells grow in colonies, or groups of stem cells, shown in the middle of this picture. Grown with the stem cells are supportive fibroblastic cells, called feeder cells, which can be seen stretched out around the colony.

In October of 2009, the California Institute for Regenerative Medicine (CIRM) announced it was awarding over $250 million to multidisciplinary research teams to develop stem cell-based therapies to treat human diseases.

Of this money, CIRM awarded $20 million for a multi-institutional grant on using human embryonic stem cells (hESCs) for vision research, with a focus on macular degeneration. Researchers at the University of California, Santa Barbara (UCSB), were part of these efforts and received $2.5 million. I was one of those researchers – I was a graduate student in the laboratory of Prof. Dennis Clegg, at UCSB, where I studied what conditions helped persuade hESCs to turn into specialized retinal cells that could be used to treat age-related macular degeneration, a leading cause of blindness. The funding from CIRM not only covered my research expenses, but it also led to the development of UCSB's state-of-the-art Center for

Stem Cell Biology and Engineering. The Center provided facilities and promoted collaborations on campus that were essential for my stem cell projects and many others' and, overall, has caused a huge and exciting push for stem cell-related research at UCSB. For more on CIRM and the research they funded at UCSB, see http://www.cirm.ca.gov/PressRelease_102809, and http://www.ia.ucsb.edu/pa/display.aspx?pkey=2122, and http://www.stemcell.ucsb.edu/. Using hESCs in vision-restoring treatments is just one of their many potential applications.

In 2013, hESCs celebrate the 15th anniversary of their discovery, and in the decade and a half since their isolation they have received a great deal of press coverage about their many potential applications, as well as debate over ethical concerns. These 15 years have also seen much progress made in better understanding these cells and their capabilities.

hESCs hold much potential for use in the field of regenerative medicine, which includes restoring damaged body tissues, organs, and limbs. They also hold great promise for serving as cellular models of human development and human diseases, to give us a better picture of these processes. However, due to ethical and application concerns, only in January of 2009 were these cells approved by the F.D.A. for first-time use in a clinical trial. This trial is discussed more later in this chapter, when Hans Keirstead's accomplishments are explored.

HISTORY OF EMBRYONIC STEM CELLS

Embryonic stem cells from other animals were successfully isolated long before the human ESCs were. Over 30 years ago, in 1981, two research groups, one headed by Dr. Gail Martin at the University of California, San Francisco, and one a collaboration between Dr. Matt Kaufman and Dr. Martin Evans at the University of Cambridge in England, independently reported the isolation of mouse embryonic stem cells (mESCs). The mESCs were isolated from early-stage mouse embryos, approximately four to six days post-fertilization (out of 21 days total for mouse gestation). It was not until 1995 that this feat was accomplished with non-human primates by Dr. James Thomson's group. And then only a few years later, in 1998, embryonic stem cells were isolated from humans, once again by a group led by Dr. Thomson, who holds a faculty appointment at the University of Wisconsin and at UCSB.

WHERE HUMAN EMBRYONIC STEM CELLS COME FROM

It is important to understand where hESCs come from in order to understand the ethical arguments that surround them, as well as their enormous, innate biological potential. After the human egg is fertilized by the sperm, the resultant single cell (called a zygote) starts dividing and multiplying. To get hESCs, cells are isolated from early-stage embryos, four or five days post-fertilization (they have not yet implanted in the uterus). At this point, the fertilized egg has developed into a hollow, fluid-filled sphere made up of approximately 150 cells. This stage is called a blastocyst.

The Blastocyst

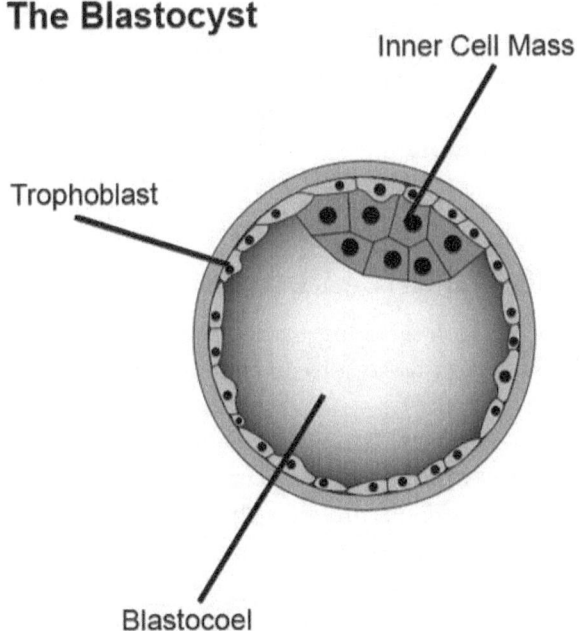

Inner Cell Mass

Trophoblast

Blastocoel

Human embryonic stem cells are isolated from early-stage embryos, four or five days after fertilization, called blastocysts. The blastocyst is a hollow sphere made of approximately 150 cells and contains three distinct areas: the trophoblast (a surrounding outer layer of cells), the blastocoel (a fluid-filled cavity), and the inner cell mass (a cluster of cells that can become the fetus). hESCs are isolated from the inner cell mass.

The blastocyst contains three distinct areas: the trophoblast, which is the surrounding outer layer of cells that later becomes the placenta; the blastocoel, which is a fluid-filled cavity within the blastocyst; and the inner cell mass, which is a cluster of cells inside the blastocyst that can become a fetus. Embryonic stem cells can be created from cells taken from the inner cell mass. Because these cells are taken from such an early stage in development, they have the ability to become cells of virtually any tissue type, which is called being "pluripotent."

The pluripotency of hESCs is probably the trait that contributes most to their enormous potential, as cellular models of human development and disease, as well as for use in regenerative medicine. This trait is what mainly separates hESCs from adult stem cells, which are called multipotent or unipotent, meaning they are only able to become a more select group of cell types. Adult stem cells are also more limited in their cellular lifespan compared to hESCs.

OVERCOMING ETHICAL CONCERNS

Because these cells are taken from human embryos, researchers have taken many careful steps to address ethical concerns. For the original creation of hESCs in 1998, blastocysts used were donated with full donor consent from *in vitro* fertilization (IVF) clinics. Additionally, many researchers use blastocysts that would have been discarded by the IVF clinic because the embryos were damaged in some way and would never develop properly. (IVF technology is discussed more in Chapter 3.)

Researchers have even found ways to isolate hESCs from embryos while leaving the donor embryo intact and potentially able to develop normally; even earlier in development than the blastocyst stage, when the developing embryo contains only eight to ten cells, researchers have shown that they can remove one of these cells and create a line of hESCs and the remaining cells will continue to function as usual. The National Academies has also developed extensive guidelines of ethical standards for researchers to follow. To read these guidelines, see

http://dels.nas.edu/bls/stemcells/guidelines.shtml.

TRANSITION TO THE CLINIC

Though many obstacles have delayed hESCs from reaching the clinic, much progress has been made in making these cells suitable for clinical use. Because of their pluripotency, one defining feature of hESCs is the regrettable tendency to create a tumor when injected into a mouse. These tumors, called teratomas, are made up of a wide variety of different cell types. To prevent the formation of tumors, all hESCs used in therapies must first be completely turned into the desired target cell type; this process of making a stem cell become a mature, more limited cell type is called differentiation. Researchers have developed methods to completely commit, or differentiate, hESCs to desired cell types, as well as many methods to confirm that they do indeed have a population that only contains these fully-committed, or differentiated, cells. Completely differentiated hESCs should not form teratomas and may be safe for therapeutic uses.

Additionally, hESCs are often grown with other supportive cells, called "feeder" cells (as is shown in the image at the beginning of this chapter). Many such feeder cells are of mouse origin. This raises concerns of non-human contaminants in hESC cultures, although it is an area of much study and many alternative methods that can create completely animal-free culture systems have been made.

Lastly, it is challenging to make patient-specific cells from hESCs, which is less of a problem when using many adult stem cells. However, this last problem, along with aforementioned ethical concerns, is being addressed with the creation of hESC-like cells from adult cells, termed induced pluripotent stem (iPS) cells, which are explored later. Human iPS cells were created in 2007 independently by two groups, one headed by Dr. James Thomson and the other by Dr. Shinya Yamanaka at Kyoto University, Japan.

Overall, hESCs have made much progress in the 15 years since their discovery, despite hurdles set before them. In March of 2009, many previous political restrictions on hESC research were removed by President Barack Obama. A few months before that, in January of 2009, the Geron Corporation had received F.D.A. approval for the first clinical trial using hESCs, specifically to treat patients with acute spinal cord injury, which is discussed in detail later in this chapter. This first clinical trial appears to have just marked the beginning for more clinical studies using these very promising stem cells. In 2010, the F.D.A. approved two more clinical trials using hESCs, led by Advanced Cell Technology (ACT) and aimed at treating two different vision conditions in patients: Stargardt's Macular Dystrophy and age-related macular degeneration (AMD). In January of 2012, preliminary data published by ACT suggests that the treatments are safe and potentially effective, though it is still too early to draw solid conclusions. In Prof. Clegg's laboratory at UCSB, I used retinal cells derived from hESCs to treat AMD, so it is very satisfying for me to see such a treatment coming to fruition and reaching the patients who desperately need it.

Further Reading

The following resources may be of interest for those wanting to learn more about human embryonic stem cells and stem cells in general:

- Teisha J. Rowland's blog "All Things Stem Cell" at http://www.AllThingsStemCell.com
- Teisha J. Rowland's blog article "Human Embryonic Stem Cells" at http://www.AllThingsStemCell.com/2009/04/human_embryonic_stem_cells/
- The National Institutes of Health's webpage "Stem Cell Information: Frequently Asked Questions (FAQs)" at http://stemcells.nih.gov/info/faqs.asp
- Teisha J. Rowland's webpage "Visual Stem Cell Glossary" at http://www.AllThingsStemCell.com/Glossary
- Teisha J. Rowland, David E. Buchholz, and Dennis O. Clegg's article "Pluripotent human stem cells for the treatment of retinal disease" at http://dx.doi.org/10.1002/jcp.22814

- Science Daily's article "Mechanisms That Allow Embryonic Stem Cells to Become Any Cell in the Human Body Identified" at http://www.sciencedaily.com/releases/2012/07/120718073729.htm
- Jef Akst's article "Supreme Court Rejects Stem Cell Case" in *The Scientist* magazine at http://www.the-scientist.com//?articles.view/articleNo/33875/title/Supreme-Court-Rejects-Stem-Cell-Case/
- Hannah Waters's article "Eye Trials Give Hope for Stem Cells" in *The Scientist* magazine at http://www.the-scientist.com/?articles.view/articleNo/31649/title/Eye-Trials-Give-Hope-for-Stem-Cells/
- For more on CIRM and the research they funded at UCSB, see http://www.cirm.ca.gov/PressRelease_102809, http://www.ia.ucsb.edu/pa/display.aspx?pkey=2122, and http://www.stemcell.ucsb.edu/.

Adult Cells Turned Stem Cell

The potential of induced pluripotent stem cells

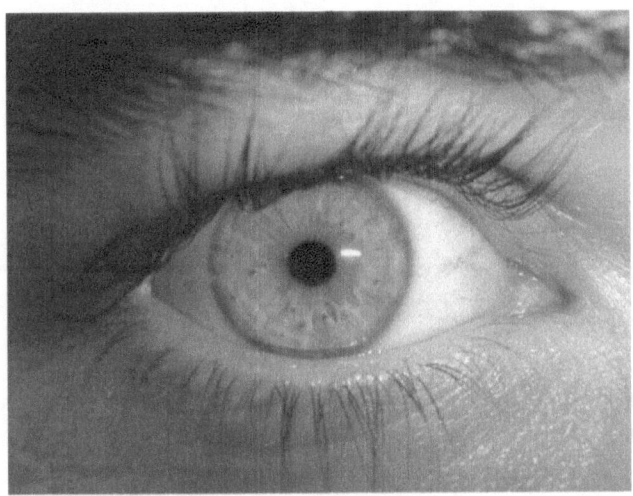

Induced pluripotent stem cells have the potential to become any cell type; recently, this ability was used to generate retinal cells that were shown to rescue vision in blind rats, suggesting these cells could also be potentially used to treat blindness in humans.

In December of 2009, two papers were published showing that induced pluripotent stem cells, also called iPS cells, could be used to rescue vision in blind rats. To read these papers, see http://dx.doi.org/10.1371/journal.pone.0008152 and http://dx.doi.org/10.1002/stem.189.

These papers were part of an international collaborative research effort led by Prof. Dennis Clegg at the University of California, Santa Barbara (UCSB), and Prof. Pete Coffey at the University College London (UCL). During this time, Clegg was my Ph.D. advisor at UCSB, and my thesis project revolved around making these studies successful.

Like all research, there were plenty of challenges, but other researchers and I at UCSB were ultimately able to make these human iPS cells differentiate, or turn into, retinal cells. Specifically, we turned the iPS cells into retinal pigment epithelium (RPE) cells.

These are essential for proper vision. Our collaborators at UCL then took these cells and transplanted them into rats that were blind due to having inherited dysfunctional RPE cells. The transplanted cells rescued the blind rats' vision; they could see.

These findings suggest that these RPE cells, derived from iPS cells, could be used to treat blindness, such as blindness caused by age-related macular degeneration (AMD), a leading cause of blindness in the Western world. This is just one of the many potential therapeutic applications of iPS cells.

WHAT ARE INDUCED PLURIPOTENT STEM CELLS?

First created in 2007, human iPS cells are a very recently created type of stem cell with great potential. iPS cells are virtually identical to human embryonic stem cells (also called hESCs), except for their origins. Both iPS cells and hESCs are pluripotent, meaning they have the potential to become any cell type in the body. However, while hESCs are created from human embryos, iPS cells are cells that were originally from adult tissues, such as skin from an adult body, but have been "reprogrammed" to a hESC-like state. The reprogramming is done by basically forcing adult cells to produce proteins that are key to the pluripotency of hESCs. The resultant cells look and act nearly identical to hESCs.

Because they are not derived from embryos, iPS cells help alleviate some of the ethical concerns surrounding the use of hESCs. Additionally, iPS cells can theoretically be patient-specific, using any adult cells from the patient to generate them. Consequently, iPS cells hold great potential for use in regenerative therapies.

To understand just how remarkable it is that iPS cells even exist, one must look at early animal development. As an embryo develops, all the cells in the embryo begin to take on different fates: The cells that will become the future internal organs are determined and are different from the predecessor cells of the skin, and so on. All of the cells in the embryo become more and more defined, and while they establish what future tissues they will be, they also define what tissues they cannot be; skin cells cannot be liver cells and vice versa. By the time the animal is an adult, most cells have very fixed identities. However, researchers have found ways to reprogram these adult cells so their identities are no longer fixed and they can theoretically become any type of cells; these reprogramming techniques led researchers to the creation of iPS cells.

HISTORY OF REPROGRAMMING CELL IDENTITY

The idea of reprogramming a cell from adult tissue into an embryonic-like cell existed long before the creation of iPS cells. In 1938, Prof. Hans Spemann (of the University of Freiburg-im-Breisgau, Germany) described his theory that any cell from an embryo that had already been developing still had the potential to create an entire adult animal. To prove his theory, Spemann used two key components: (1) a fertilized egg that had had its nucleus removed (what is called "enucleated"), and (2) a nucleus from an older cell. (The nucleus is the part of a cell that contains the genetic material, or DNA.) Spemann conducted his studies in salamander eggs, which are large and easy to manipulate. He would take a newly fertilized salamander egg, remove its nucleus, and put the nucleus from an older salamander embryo into the enucleated egg. Somehow, the older nucleus could still give rise to an entire adult salamander in this way.

In the early 1950s, Profs. Robert Briggs and Thomas King (at the Fox Chase Cancer Center, in Philadelphia, Pennsylvania) repeated Spemann's experiments using leopard frogs and had success using a nucleus from an even later point in embryonic development.

In essence, these studies suggested that a newly fertilized, enucleated egg has the ability to "reprogram" a much older donor nucleus, making the resultant cell behave as though it were at a very early embryonic state. For years it was unclear whether the nucleus from a completely mature, adult cell could be reprogrammed because conflicting results were published by different research groups.

SOMATIC CELL NUCLEAR TRANSFER AND CLONING

Although the studies done by Spemann, Briggs, and King used nuclei from embryos, their results formed the basis for somatic cell nuclear transfer (SCNT). SCNT is a technique wherein the nucleus from a somatic cell (an adult cell that is not a sperm or an egg) is implanted into an enucleated egg which can be implanted into, and develop, in a surrogate mother, and potentially become an adult organism. The resultant organism is a clone of the animal that donated the nucleus. The first widely-accepted successful use of SCNT came with the creation of the sheep Dolly in 1997, the first cloned animal from an adult. Since then, several other animals have been cloned, though many problems still remain. (More about other applications of SCNT, specifically therapeutic cloning, is discussed later in this chapter.)

CREATION OF INDUCED PLURIPOTENT STEM CELLS

With the living evidence of Dolly and other animals cloned from adult cells, the idea that an adult somatic cell could become a reprogrammed embryonic-like cell regained a spotlight in the scientific community. The creation of iPS cells began by studying proteins not only uniquely created by embryonic stem cells, but proteins known to be functionally important in creating the unique properties of these cells. In 2006, Shinya Yamanaka's group at the Institute for Frontier Medical Sciences at Kyoto University made the first iPS cells by applying this knowledge in mouse cells. They made adult mouse cells become essentially mouse embryonic stem cells, in appearance and function, by forcing the adult cells to produce four key protein factors. The DNA of these factors was introduced into the cells using a retroviral system (a class of viruses similar to HIV in function but not in severity), incorporated into the cells' genome, and consequently translated into functioning protein by the host cell.

The same principles applied to the creation of human iPS cells, reported only a year later concurrently, though independently, by the laboratories of Yamanaka and Prof. James Thomson (who holds a faculty appointment at the University of Wisconsin and at UCSB). Yamanaka's group used human adult skin cells and induced them to become iPS cells by having them produce the same protein factors that the mouse iPS cells had.

Thomson's group created human iPS cells using fetal lung cells and foreskin cells with two of the same factors as Yamanaka, along with two different factors. The two protein factors the groups had in common appear to be quite crucial for the creation of iPS cells. Even though different cell types were used as the initial starting materials, and they were made to produce different sets of proteins, both groups identified and isolated cells nearly identical to human embryonic stem cells, and did so in the same timeframe.

hESCS VS. iPS CELLS

Though iPS cells and hESCs are both pluripotent and virtually infinite in supply, there are distinct advantages and disadvantages associated with each cell type. As they are created from adult cells, iPS cells overcome many ethical concerns associated with hESCs and may be patient-specific, but may have life spans dependent upon donor cell age. Additionally, the originally generated iPS cells contained DNA randomly inserted into the genome from the retrovirus system, although researchers have since made new forms of iPS cells using non-integrating systems and have also improved delivery of the reprogramming factors to the cells.

Though issues are still being addressed to ensure their general safe use in regenerative medicine, iPS cells are quickly overcoming these hurdles to make their way to patients in clinical trials. Recently, in February 2013, it was announced that Prof. Masayo Takahashi and colleagues at the Center for Developmental Biology in Kobe, Japan, were almost ready to use RPE cells derived from iPS cells to treat AMD – the same goal I was pursuing as a graduate student at UCSB in Prof. Dennis Clegg's laboratory. Prof. Takahashi's treatment, which has been approved by an institutional review board in Kobe, will likely be the first treatment to use iPS cells in a human trial if the results of the preclinical safety studies are satisfactory. For more information on this, see http://www.the-scientist.com//?articles.view/articleNo/34396/title/Stem-Cell-Trial-Nearly-Approved/. Only with additional research, requiring continual funding, can additional promising treatments utilizing iPS cells make it to the patients who need them, such as a treatment like this one for AMD that uses patient-specific, specialized retinal cells.

Further Reading

The following resources may be of interest for those wanting to learn more about induced pluripotent stem cells and stem cells in general:

- Teisha J. Rowland's blog "All Things Stem Cell" at http://www.AllThingsStemCell.com
- Teisha J. Rowland's blog article "Induced Pluripotent Stem Cells: A New Stem Cell Line with a Long History" at http://www.allthingsstemcell.com/2009/06/induced-pluripotent-stem-cells-a-new-stem-cell-line-with-a-long-history/
- The National Institutes of Health's webpage "Stem Cell Information: Frequently Asked Questions (FAQs)" at http://stemcells.nih.gov/info/faqs.asp
- Teisha J. Rowland's webpage "Visual Stem Cell Glossary" at http://www.AllThingsStemCell.com/Glossary
- Dan Cossins's article "Stem Cells Not Rejected" in *The Scientist* magazine at http://www.the-scientist.com//?articles.view/articleNo/34129/title/Stem-Cells-Not-Rejected/
- Jef Akst's article "Stem Cell Trial Nearly Approved" in *The Scientist* magazine at http://www.the-scientist.com//?articles.view/articleNo/34396/title/Stem-Cell-Trial-Nearly-Approved/
- Hayley Dunning's article "Creating Sperm from Skin" in *The Scientist* magazine at http://www.the-scientist.com/?articles.view/articleNo/32565/title/Creating-Sperm-from-Skin/
- Amanda-Jayne Carr et al.'s article "Protective Effects of Human iPS-Derived Retinal Pigment Epithelium Cell Transplantation in the Retinal Dystrophic Rat" at http://dx.doi.org/10.1371/journal.pone.0008152
- David E. Buchholz et al.'s article "Derivation of Functional Retinal Pigmented Epithelium from Induced Pluripotent Stem Cells" at http://dx.doi.org/10.1002/stem.189
- Teisha J. Rowland, David E. Buchholz, and Dennis O. Clegg's article "Pluripotent human stem cells for the treatment of retinal disease" at http://dx.doi.org/10.1002/jcp.22814

Direct Reprogramming

Changing a cell's identity

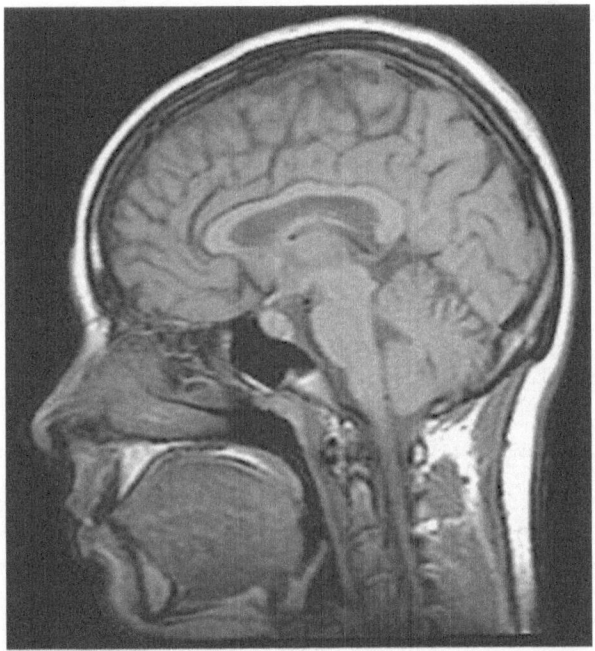

Recently researchers found that cells from the tails of mice could be turned into neurons (the core cells of the brain). This was done through a process called "direct reprogramming;" the tail cells were forced to produce proteins essential for the identity of neurons, and consequently looked and functioned like neurons.

The idea that a small piece of skin could be taken from a person and turned into a kidney useful for a transplant sounds like science fiction. But in the last few decades such amazing progress has been made in the stem cell field as to suggest that someday in the not-so-far future, this may be a reality.

FROM iPSCS TO DIRECT REPROGRAMMING

Stem cells hold great potential for treating diseases and regenerating tissues. One key reason for this is that stem cells can turn into, or differentiate into, many different types of cells. Equally valuable is their ability to be patient-specific; cells can be taken from a patient and be used to treat that same patient, bypassing potential immune rejection. Possibly embodying these two traits more than any other stem cells, human induced pluripotent stem cells (iPSCs) can become any cell type in the body and be made from a patient's own cells. Human iPSCs were first created in 2007 and discussed previously in this chapter.

Although iPSCs are not yet being used clinically in patients, the theoretical procedure for making iPSCs for treatments is time-consuming. Adult cells must first be obtained, then "reprogrammed" into an embryonic-like state where they can potentially turn into any cell type, and lastly coaxed into becoming the final desired cell type. "Reprogramming" involves forcing the adult cells to produce factors that are normally only made in embryonic cells. These factors (called transcription factors) change the proteins the adult cell produces so that it starts to make embryonic proteins, eventually making it look and function like an embryonic stem cell (embryonic stem cells were discussed earlier in this chapter). Just as embryonic stem cells can become any cell type, so can their mimics, the induced pluripotent stem cells.

Researchers over the last few decades have been investigating ways to bypass the intermediate embryonic-like state and reprogram adult cells directly into the final desired cell type, an approach called "direct reprogramming." Their work has shown that this is clearly possible, although the underlying mechanisms remain poorly characterized. Most of these studies have been done using a few cell types: muscle, blood, pancreatic cells, and neurons.

MAKING MUSCLE WITH MyoD

In the late 1980s, researchers found that many different types of cells could be turned into muscle cells, in the first direct reprogramming experiments done with transcription factors. In 1987, Dr. Harold M. Weintraub and colleagues at the Department of Genetics Fred Hutchinson Cancer Research Center in Seattle, Washington, found that when fibroblasts (cells in connective tissue that are important for wound healing) were forced to produce a transcription factor the researchers named MyoD, the fibroblasts became muscle cells. MyoD was later found to be important in turning cells into muscle during normal development.

Because it worked so well in fibroblasts, MyoD was tested for its ability to reprogram other cell types into muscle. Indeed, in the early 1990s, other groups found that MyoD could in fact make skin cells, cells from cartilage, and cells from the retina, among others, all become muscle. However, not all cells were easily persuaded to change their fates; liver cells couldn't become muscle. Unlike many scientists who have made ground-breaking discoveries, Dr. Weintraub lived to see his discovered transcription factor, MyoD, recognized as an important factor before he died of cancer in 1995, at the age of 49.

DIRECTLY REPROGRAMMING THE BLOOD SYSTEM

The 1990s saw the arrival of many more important transcription factors to the direct reprogramming scene. One biological system that had its reprogramming abilities heavily explored was blood. A key group of stem cells, called hematopoietic stem cells (HSCs), is responsible for creating all of the different blood cells the body needs to combat infections and deliver oxygen to its organs, among other essential duties. The creation process is called hematopoiesis (which literally means "to make blood"), and is shown in the image below. During hematopoiesis, a HSC first differentiates into either a myeloid or a lymphoid progenitor cell. Each progenitor cell can then become one of two major blood groups. The group descended from myeloid progenitors (shown on the left in the image below) contains white blood cells (monocytes, eosinophils, and macrophages), red blood cells (erythrocytes), and platelets (thrombocytes). The second group, descended from lymphoid progenitors (and shown on the right), is made up of lymphocytes (including B-cells and T-cells).

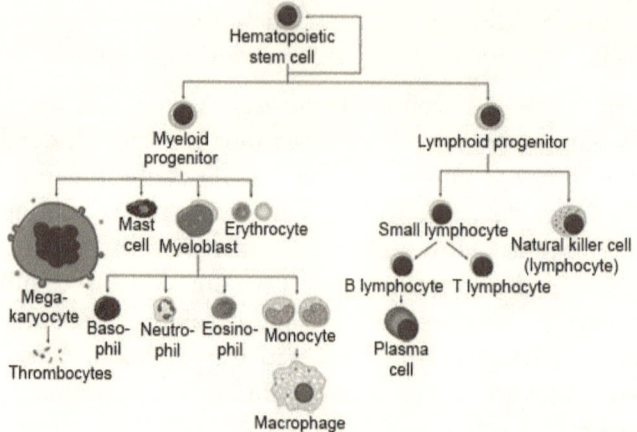

Hematopoietic
stem cell

Myeloid
progenitor

Lymphoid progenitor

Mast
cell   Myeloblast

Erythrocyte

Small lymphocyte

Natural killer cell
(lymphocyte)

B lymphocyte T lymphocyte

Mega-
karyocyte

Baso-   Neutro- Eosino- Monocyte
phil   phil   phil

Plasma
cell

Thrombocytes

Macrophage

During hematopoiesis, hematopoietic stem cells first become either myeloid progenitors or lymphoid progenitors. From there, the myeloid progenitors become white blood cells (monocytes, eosinophils, and macrophages), red blood cells (erythrocytes), platelets (thrombocytes), or other blood cell types, while lymphoid progenitors become lymphocytes (including B-cells and T-cells).

In 1995, Dr. Thomas Graf's group at the European Molecular Biology Laboratory in Heidelberg, Germany, found that changing the amount of a transcription factor called GATA-1 that is present in white blood cells (specifically monocytes) could make these cells turn into other kinds of blood cells from the same group: other white blood cells (eosinophils), red blood cells, or platelets, which are essential for blood clotting. This finding is depicted in the image below. This correlated with the normal role of GATA-1 during hematopoiesis; different amounts of GATA-1 are present in these different blood cell types, and the presence or absence of GATA-1 controls what fate these cells usually have. While this finding might not seem as impressive as being able to turn many different types of cells into muscle, these experiments were done in the well-characterized blood system, so that even small cell-fate changes could be better understood.

Over the last decade, Dr. Graf's group has continued to better understand reprogramming in the blood system. In 2004, he and his researchers found that when lymphocytes (specifically B lymphocytes) were forced to produce key transcription factors (a group called C/EBPs), these lymphocytes could become white blood cells (specifically macrophages). In other words, they were able to make one blood cell become a blood cell type from the other major blood cell group. This makes sense with the normal role of C/EBPs in hematopoiesis; they are necessary for monocytes to develop into macrophages.

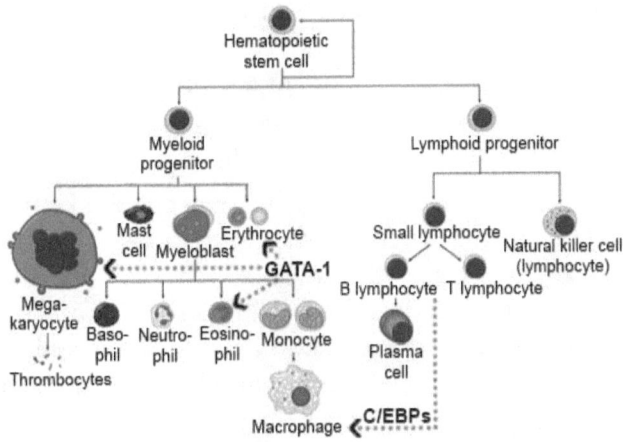

Changing the level of the transcription factor GATA-1 in monocytes can make them differentiate into white blood cells (eosinophils), red blood cells (erythrocytes), or megakaryocytes, which become platelets (thrombocytes). It's also been found that forcing B lymphocytes to make C/EBP transcription factors causing these cells to turn into macrophages, a blood cell type from the other major blood cell group.

Looking at the bigger picture, this 2004 report is significant in that it was the first report showing that cells completely committed to their fates, or fully differentiated, could be reprogrammed using transcription factors; it was the first report of what's known as transdifferentiation. (HSCs, which are stem cells that have been used to treat leukemia, lymphomas, and HIV, are discussed in greater depth in the next section of this chapter.)

DIRECTLY REPROGRAMMING IN THE PANCREAS

While it seemed easy enough to make cells become muscle, or make blood cells become other kinds of blood cells, it took a great amount of effort to make such direct reprogramming possible in other parts of the body. It was found to be possible, and in 2008 Dr. Douglas A. Melton's group at the Harvard Stem Cell Institute, at Harvard University, in Cambridge, Massachusetts, reported that they had made one kind of cell in the pancreas, specifically exocrine cells, become another, functionally different type of pancreatic cell, beta-cells. While exocrine cells secrete digestive enzymes used by the intestine, beta-cells are important for producing insulin. Since diabetes can be caused by a lack of insulin, it is very appealing to be able to create more insulin-producing cells.

Dr. Melton's group screened over 1,000 transcription factors to find which could make the exocrine cells turn into beta-cells. Based on how the exocrine cells responded to the factors, the group narrowed down their list to a few key factors; eventually three used together made them into cells like beta-cells. Not only did they look like beta-cells, but they also produced proteins made by beta-cells, and generated insulin. This discovery renewed reprogramming efforts in other cell types, and gave new hope to diabetes research.

FROM THE TAIL TO THE BRAIN

With the exception of MyoD and muscle, it was still unclear whether a cell could be reprogrammed into a very distantly related cell type. This brings us to a more recent direct reprogramming breakthrough. In January 2010, Dr. Marius Wernig's group at the Stanford University School of Medicine showed that fibroblasts from the tails of mice could be turned into nerve cells, or neurons. After screening 19 candidate factors, the researchers found three (all of them normally involved in neuronal development) that turned the fibroblasts into functional neurons.

While it is clearly possible to directly reprogram one cell type into another, it is still very unclear how this happens. For instance, it is not known whether a "directly reprogrammed" cell actually directly becomes the final cell type seen, or whether it undergoes a transition state before settling on the final cell type. Aside from using transcription factors, several other methods of reprogramming are being pursued (such as using microRNAs or chemicals, or altering chromatin states). It may turn out that the most effective approach is to combine these different methods.

While direct reprogramming results are quite promising, they are still at a very early stage; most of the studies reported have been in mice or mouse cells and must be repeated using human cells before they can be considered for use in humans. However, if such reprogramming studies continue to be successful, they may play a key role in creating patient-specific cells for treatments and regenerative medicine, as well as ideal cells for studying development and diseases that occur when cells lose their identity, such as cancer.

Further Reading

The following resources may be of interest for those wanting to learn more about direct reprogramming and stem cells in general:

- Teisha J. Rowland's blog "All Things Stem Cell" at http://www.AllThingsStemCell.com
- Teisha J. Rowland's blog article "Direct Reprogramming: Turning One Cell Directly Into Another" at http://www.allthingsstemcell.com/2010/02/direct-reprogramming-turning-one-cell-directly-into-another/
- The National Institutes of Health's webpage "Stem Cell Information: Frequently Asked Questions (FAQs)" at http://stemcells.nih.gov/info/faqs.asp
- Teisha J. Rowland's webpage "Visual Stem Cell Glossary" at http://www.AllThingsStemCell.com/Glossary
- Thomas Vierbuchen et al.'s article "Direct conversion of fibroblasts to functional neurons by defined factors" in Nature at http://dx.doi.org/10.1038/nature08797

Stem Cells That Treat Leukemia, Lymphoma, and HIV

Hematopoietic stem cells can become all different blood cell types

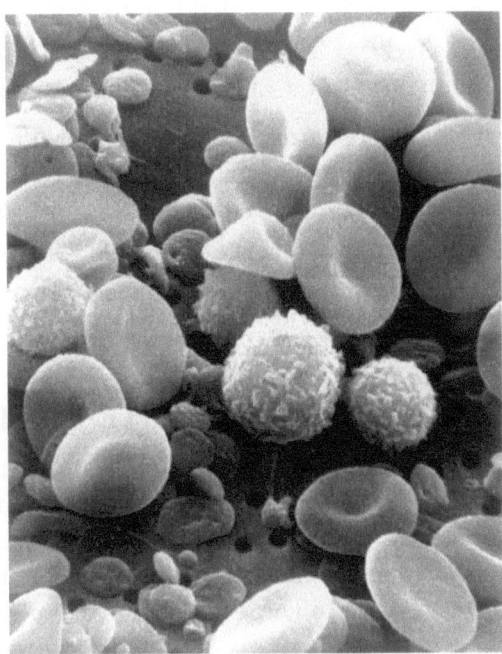

Hematopoietic stem cells have the ability to become all the different types of blood cells; above are a variety of blood cells found in a normal sample of circulating blood.

Hematopoietic stem cells (HSCs), cells that turn into blood cells, have been heavily researched for decades, and have been successful in treating blood-related diseases in humans, such as leukemia and lymphoma, for much of this time. Recently, they were even successfully used in treating a patient with both leukemia and HIV.

DISCOVERY OF HEMATOPOIETIC STEM CELLS

Like many significant findings in science, their discovery was not made in search for such cures, but it was stumbled upon while dealing with another serious medical issue of the time: radiation exposure. During and immediately following World War II, scientists tried desperately to treat people who had been exposed to lethal doses of radiation. Transplants of spleen and bone marrow tissue were found to rescue these victims. It was later found that HSCs present in these tissues were behind these healing properties; researchers discovered this by lethally irradiating rats and mice and similarly rescuing them with transplants of different types of cells.

The formation of the National Marrow Donor Program in the 1980s greatly increased the availability of these cells for research. For more on the National Marrow Donor Program, see http://www.marrow.org/. HSCs have been successfully used clinically in humans since the 1950s, and to this day they are one of the few adult stem cells widely used in clinical therapies.

HEMATOPOIETIC STEM CELL POTENTIAL

HSCs are able to become, or "differentiate" into, all of the cells in the hematopoietic system. The hematopoietic system includes all the different kinds of blood cells, from myeloid elements (including red blood cells, white blood cells, and platelets) to lymphatic elements (such as T-Cells). These tissues are some of the most rapidly dividing cells in the body, and as such are generally the first cells hurt by radiation. For this reason, therapies using HSC transplants allow cells that are generally the most damaged by exposure to be replenished.

It is possible to isolate HSCs from a wide variety of both adult and fetal human tissues, including adult bone marrow, fetal liver, spleen, and thymus, umbilical chords, and blood. In recent years, there has been a great shift toward obtaining HSCs mainly from blood, using a much simpler and less controversial procedure. However, collecting a large enough number of HSCs for successful transplants remains a challenge.

It's now not only better understood how HSCs from a donor can save a lethally irradiated recipient, but also how HSCs can be used in many other medical applications. HSCs are now being used to treat cancers of the hematopoietic system (leukemia and lymphomas), replenish cells lost to high doses of chemotherapy, and fight against autoimmune diseases, in addition to other medical applications.

STEM CELLS THAT CAN CURE AIDS

Recently, an amazingly successful application of HSC therapy was used to treat a patient with both leukemia and human immunodeficiency virus (HIV). Drs. Eckhard Thiel, Wolf. K Hofmann, and colleagues (at the Department of Hematology, Oncology, and Transfusion Medicine, Charité Universitätsmedizin, Berlin, Germany) gave the patient special HSC transplants; amazingly, the patient had remission of both conditions after the treatment. As mentioned, HSCs have been used in transplants to rescue patients with leukemia, but this method had not previously been as successful for treating HIV, the virus that causes acquired immunodeficiency syndrome (AIDS).

To understand how the HSC transplant may have cured AIDS in this patient, it's important to understand how HIV invades the body on a molecular level. Once in the body, HIV primarily attacks the immune system, including the T cells. However, some individuals have T cells that are naturally resistant to HIV infection. Over a decade ago, this resistance was found to be due to a mutation in a protein that is normally on the surface of T cells (called chemokine receptor 5 [CCR5]). This protein is thought to normally be involved in causing an immune response to infection, though its full function is still not understood. HIV normally must interact with CCR5 in order to gain entry into the T cell. However, in individuals with a mutated version of CCR5, the protein is not functional and consequently does not allow HIV to enter the T cells.

The CCR5 protein has been greatly studied since its discovery, as it has been looked upon as a potential cure for HIV. Approximately 5 percent to 14 percent of Europeans have the mutated version of this protein, but it is absent in African, East Asian, and Native American populations. The mutation is thought to have arisen over 5,000 years ago. Surprisingly, individuals with two mutated CCR5 copies have no apparent health problems. Anti-HIV drugs have been created that bind the normal CCR5 protein, attempting to block HIV entry into the T cells.

The recent success story that applied this knowledge of CCR5 and MSCs, mentioned above, was reported in February 2009. A 40-year-old Caucasian patient with leukemia and HIV underwent HSC transplants using blood from a donor with (two copies of) the mutated CCR5 protein. (This means the patient's body underwent irradiation treatment to kill the existing HSCs, and these were then replaced by the donor's HSCs.) Over 20 months later, not only did the leukemia go into complete remission, but the patient had no HIV detectable in his body either. As happens in treating leukemia, the donor HSCs replaced his irradiated immune system, but in this particular case this also meant that the normal CCR5 proteins on his T cells were replaced by CCR5-mutated, HIV-resistant T cells. To read the actual paper, see

http://dx.doi.org/10.1056/NEJMoa0802905.

While this is quite the success story, a stem cell transplant like this, which involves risky irradiation of the patient, is still usually a less safe prospect than using well-developed HIV therapy drugs, though this could change with additional research. This treatment should be repeated in other patients to confirm the validity of the technique, and it must also be taken into consideration that some HIV strains do not access T cells using CCR5, but use other surface proteins instead.

This success, however, highlights the fact that while HSCs have been clinically used and studied to a large extent relative to other human stem cells, a great deal of work remains to be done to fully understand their molecular interactions and potential in medicine, which can only be accomplished with continual funding and support.

Further Reading

The following resources may be of interest for those wanting to learn more about hematopoietic stem cells and stem cells in general:

- Teisha J. Rowland's blog "All Things Stem Cell" at
  http://www.AllThingsStemCell.com
- Teisha J. Rowland's blog article "Hematopoietic Stem Cells: A Long History in Brief" at

http://www.allthingsstemcell.com/2009/02/hematopoietic-stem-cells/

- Teisha J. Rowland's blog article "Potential of Stem Cells to Cure HIV" at http://www.allthingsstemcell.com/2009/03/stem-cells-cure-hiv/
- The National Institutes of Health's webpage "5. Hematopoietic Stem Cells" at http://stemcells.nih.gov/info/scireport/chapter5.asp
- The National Institutes of Health's webpage "Stem Cell Information: Frequently Asked Questions (FAQs)" at http://stemcells.nih.gov/info/faqs.asp
- Teisha J. Rowland's webpage "Visual Stem Cell Glossary" at http://www.AllThingsStemCell.com/Glossary
- Gero Hütter et al.'s article "Long-Term Control of HIV by CCR5 Delta32/Delta/32 Stem-Cell Transplantation" at http://dx.doi.org/10.1056/NEJMoa0802905
- *Science Daily's* article "Stem Cell Therapy Shows Promise in Fight Against HIV" at http://www.sciencedaily.com/releases/2012/05/120502092042.htm
- David T. Harris's article "Opinion: Younger is Better" in *The Scientist* magazine at http://www.the-scientist.com/?articles.view/articleNo/32483/title/Opinion--Younger-Is-Better/

Stem Cells in Your Teeth and Other Mesenchymal Stem Cells

Powerful potential in an unlikely location

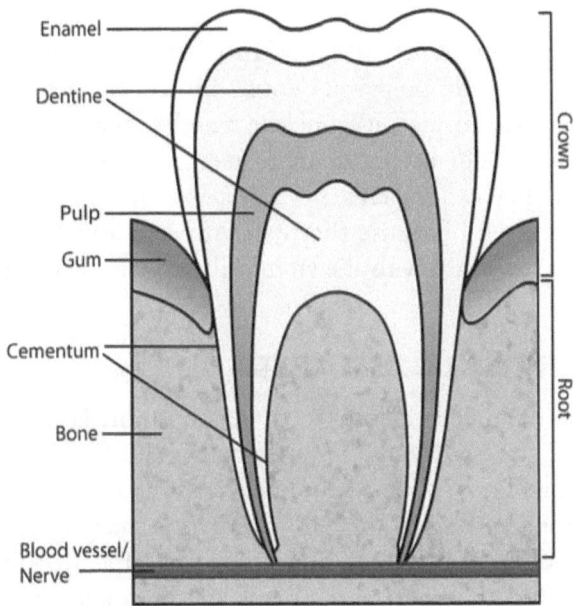

Potentially powerful stem cells have been found to reside in human teeth. These stem cells are easy to obtain when children lose their baby teeth and have many potential medical applications, such as repairing bone tissue.

It's been slowly becoming apparent that stem cells reside just about everywhere inside the human body. Even in the seemingly unlikely places, such as teeth.

DENTAL PULP STEM CELLS

Because children lose their baby teeth as they grow up, the stem cells in teeth are one of few types of stem cells that are naturally molted, or exfoliated. Consequently, because these stem cells can be harvested noninvasively, without harming the patient, they are quite appealing for medical applications. These stem cells reside in the dental pulp of the tooth, which is the living tissue at the center of the tooth, and so were aptly named dental pulp stem cells (DPSCs) by their discoverers in 2000. In addition to being easy to come by, DPSCs also have the potential to become many different types of cells.

As you might guess, DPSCs can produce some of the primary components of teeth: pulp and dentin. However, they cannot grow entire new teeth. In fact, it's unclear whether they can even create the other two major components in teeth, enamel and cementum. But, even if they cannot create whole teeth by themselves, DPSCs are still quite useful because they can turn into a wide range of cell types, which has to do with the stem cell category they belong to.

MESENCHYMAL STEM CELLS

Dental pulp stem cells belong to the large umbrella family of stem cells called mesenchymal stem cells (MSC). The term "mesenchymal" may sound daunting, but it just refers to a part of the developing embryo. The early embryo is made up of three cell layers: endoderm, mesoderm, and ectoderm. Each of these layers becomes different tissues and organs of the adult animal's body, together making a whole. The mesoderm layer becomes support structures throughout the body, such as bone, cartilage, connective tissue, smooth muscle, adipose tissue, lymph nodes, and blood cells. MSCs are mostly derived from mesenchymal tissues, which largely originate from this mesoderm layer.

In addition to teeth, MSCs have also been harvested from many different tissues in the human body (bone, cartilage, muscle, umbilical cords, adipose tissue, blood, amniotic fluids, and some fetal tissues). Since the 1960s and 1970s it's been known that there are MSCs in bone marrow. (However, because other stem cells were already discovered in bone marrow, specifically hematopoietic stem cells, MSCs were not widely accepted as a stem cell type until the late 1990s.) To date, MSCs from bone marrow and adipose tissue have been the most studied, as they are relatively easy to harvest in large numbers.

Since their relatively recent discovery and addition to the MSC family, dental pulp stem cells have been greatly characterized and found to literally have great potential. To understand what cell types DPSCs can become, it's important to know what cell types MSCs in general can turn into. Although MSCs have the potential to become multiple different types of cells (they're "multipotent"), the three key cell types all MSCs can become are osteocytes (bone cells), chondrocytes (cartilage cells), and adipocytes (fat cells). Since they are MSCs, DPSCs can become these three cell types. However, different MSCs can turn into different, additional cell types depending on the tissue they were harvested from. For example, some MSCs, including DPSCs, can also become muscle cells.

Interestingly, DPSCs actually display some properties of another group of stem cells as well. Most researchers classify them as MSCs, but DPSCs are also similar to neural crest stem cells (NCSCs) (the term "neural crest" again applies to a specific part of the developing embryo). For example, DPSCs can become melanocytes, which are not a mesenchymal cell type. Clearly, further characterization of DPSCs is desirable as even their basic classification remains unclear.

APPLICATIONS

Because they can be obtained in a completely noninvasive manner and have relatively broad differentiation capabilities, dental pulp stem cells have great potential to be used in medical applications, such as healing wounds and regenerating tissues. MSCs in general have been significantly pursued for these applications, especially since MSCs have been found to inhibit inflammation and immune responses. A major barrier, which has prevented more widespread use of MSCs (including DPSCs) in the medical field, has been how they are harvested. When the groups of cells are isolated from tissue they can be rather heterogeneous, containing the desired stem cells but also many other cell types. Also, the cells can vary from donor to donor. However, great efforts are being made to standardize isolation procedures to bring experiments from the laboratory to clinical utility.

In early 2010, dental pulp stem cells were successfully used in their first human clinical trial. Specifically, studies reported that DPSCs could be used to repair bone tissue. For information on one such clinical study, see

http://www.ecmjournal.org/journal/papers/vol018/vol018a07.php.
Researchers have been investigating whether DPSCs may also be used to safely repair or regenerate other tissues and organs. In rats, DPSCs have also been shown to reduce myocardial infarctions (heart attacks). To read about this study, see

http://dx.doi.org/10.1634/stemcells.2007-0484.

Clearly DPSCs hold much potential, and biotechnology companies such as the National Dental Pulp Laboratory in Massachusetts have already jumped at the opportunity. They're offering to cryogenically freeze the DPSCs from baby teeth so that the DPSCs can potentially be used later in life if they're needed (just as many companies offer to do for umbilical cords, which contain mesenchymal and hematopoietic stem cells). In other words, the tooth fairy may have just found a way to make a tidy profit from all those baby teeth.

Further Reading

The following resources may be of interest for those wanting to learn more about dental pulp stem cells, mesenchymal stem cells, and stem cells in general:

- Teisha J. Rowland's blog "All Things Stem Cell" at http://www.AllThingsStemCell.com
- Teisha J. Rowland's blog article "Dental Pulp Stem Cells: Noninvasively Obtained Stem Cells" at http://www.allthingsstemcell.com/2009/02/dental-pulp-stem-cells/
- Teisha J. Rowland's blog article "Mesenchymal Stem Cells: A Diverse Family, Large and Still Growing" at http://www.allthingsstemcell.com/2009/03/mesenchymal-stem-cells/
- The National Dental Pulp Laboratory's website at http://www.ndpl.net/

- The National Institutes of Health's webpage "Stem Cell Information: Frequently Asked Questions (FAQs)" at http://stemcells.nih.gov/info/faqs.asp
- Teisha J. Rowland's webpage "Visual Stem Cell Glossary" at http://www.AllThingsStemCell.com/Glossary
- Carolina Gandia et al.'s article "Human Dental Pulp Stem Cells Improve Left Ventricular Function, Induce Angiogenesis, and Reduce Infarct Size in Rats with Acute Myocardial Infarction" at http://dx.doi.org/10.1634/stemcells.2007-0484
- R. d'Aquino et al.'s article "Human mandible bone defect repair by the grafting of dental pulp stem/progenitor stem cells and collagen sponge biocomplexes" at http://www.ecmjournal.org/journal/papers/vol018/vol018a07.php

Stem Cells in Menstrual Blood

They can even become neurons

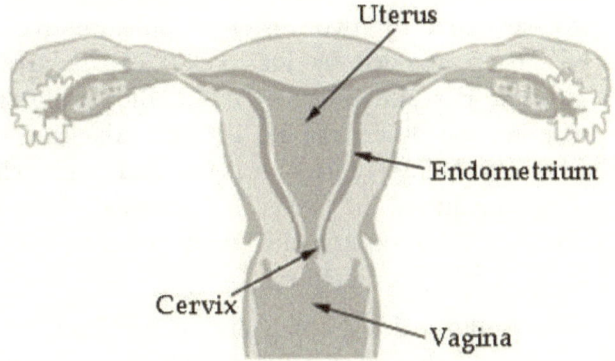

In 2007, stem cells were found in menstrual blood, which is made up of shed endometrium. These stem cells hold great therapeutic potential and already their ability to treat several diseases is being thoroughly explored.

One of the major barriers preventing the widespread use of stem cells in therapeutic clinics is the great difficulty of collecting a large number of stem cells from any single person. But this isn't because our bodies don't have many stem cells; they are, in fact, present in most tissues. Rather, there are few tissues that freely shed their cells on a regular basis and in great quantity. Consequently, stem cells must often be harvested using an invasive procedure. For example, one of the least invasive procedures is using a needle and syringe to withdraw stem cells from bone marrow, and only so many stem cells can be removed this way.

ENDOMETRIAL REGENERATIVE CELLS

However, a relatively recently discovered addition to the stem cell family offers a way to overcome this hurdle. Stem cells have been found to reside in menstrual blood, presenting a possible noninvasive manner to obtain large numbers of stem cells from a single patient. These stem cells hold great therapeutic potential for treating diseases and regenerating lost tissues, although much still remains to be learned about them.

Stem cells residing in menstrual blood, referred to as endometrial regenerative cells (ERCs), were discovered in 2007 and independently reported by two different collaborative research groups. (One group was led by Dr. Neil Riordan, president of MediStem Laboratories and director of the Bio-Communications Research Institute; the other was led by Julie Allickson of Cryo-Cell). In addition to being regularly available and freely released by the body in large quantities, the ERCs also potentially overcome the problem of immune rejection in many female patients, as they could use their own stem cells for therapies.

Such incredible findings don't come out of nowhere; researchers had already suspected that stem cells may be present in menstrual blood because stem cells were previously found in a layer of cells that line the wall of the uterus. On a monthly basis, this layer, called the endometrium, becomes thicker and increases the number of blood vessels and small glands within it to create ideal conditions for nurturing an embryo. However, if embryo implantation does not occur, the early endometrium lining is broken down and shed: menstruation.

Because the endometrium is constantly growing, constantly being broken down and rebuilt, it's an ideal tissue to investigate for the presence of stem cells; stem cells are masters at creating tissues. The menstrual blood excreted monthly is made up of both blood cells and cells from the endometrium layer. It was first reported in 2004 that stem cells reside within the intact endometrium lining of the uterus. Researchers therefore thought it likely that stem cells could also be found in the shed endometrium. Surprisingly, the stem cells discovered in menstrual blood, ERCs, appear to be different from stem cells living in the intact endometrium.

WHERE DO THEY FIT IN TO THE STEM CELL FAMILY?

What's surprising is that while stem cells from the intact endometrium belong to a large group of stem cells called mesenchymal stem cells (MSCs) (which includes dental pulp stem cells and many other types of stem cells that were discussed in the previous section in this chapter), ERCs do not belong to the MSC group; they make different proteins and can become different types of cells. More specifically, MSCs create a known, large set of proteins. The ERCs (which actually physically appear similar to MSCs) produce some, but not all, of the proteins in the established MSC set. Additionally, ERCs were found to be able to turn into many more types of cells than the stem cells from the intact endometrium. For example, ERCs can become neurons, or brain cells, as well as muscle, fat cells, bone, cartilage, liver cells, and other cell types. Overall, ERCs were determined to be functionally distinct from endometrium MSCs. Although ERCs do not appear to be MSCs, the stem cell category ERCs best fall into remains unclear, as do several other answers concerning their stem cell identity.

Again surprisingly, ERCs do share some characteristics with embryonic stem cells, producing some proteins that are normally made by embryonic stem cells, and that are quite unusual for cells in the adult body to make. (For more on embryonic stem cells, see their dedicated section in this chapter.) The presence of embryonic stem cell proteins in ERCs, combined with the ability of ERCs to differentiate into a large number of cell types, led one research group to propose that ERCs may be able to become any desired cell type, although other researchers report that their stem cell potential is more limited; overall, much remains to be determined. If the ERCs hold the great promise that some researchers believe they do, they may be a vital resource for future cell-based therapies.

Some reported characterization discrepancies of the ERCs between research groups have led to speculation that there is variability in the quality of ERCs isolated. Since there was not a standard method of isolation in place for these cells, there could be great variability in the actual cells isolated, depending on the specifics of the purification method used. Additionally, the quality of ERCs isolated could depend on the individual donors, possibly being related to age or other factors. To complicate the identity of these stem cells even further, it has been suggested that there may be multiple different types of stem cells living in the menstrual blood, not just the ERCs; it may be difficult to be sure of exactly what cell type is being tested.

In addition to questions about their identity, the origin of these stem cells is also still very much up for debate. Some theorize that the ERCs are endometrium stem cells that were shed, though, as discussed above, because ERCs are rather distinct from the intact endometrium stem cells it makes it quite unclear. Other researchers hypothesize that the ERCs are from endometrial glands, as many glands are present in menstrual blood. With a better understanding of the origins of these cells in the body, and potential variability between donors, it will be easier to properly isolate the ERCs for use in down-stream clinical applications.

MUCH TO BE FIGURED OUT, BUT ALREADY GOING TO TRIALS

Though many questions remain to be answered before these newly discovered cells can be accurately characterized, a great amount of interest in using menstrual blood stem cells for regenerative therapies has already materialized. Researchers have been looking into using ERCs for treating neurodegenerative and cardiovascular diseases and salvaging limbs, among several other applications. Preliminary clinical trials for treating multiple sclerosis in humans using ERCs have already yielded some promising results, and the FDA has granted approval for these cells to be used in other human clinical trials. Several industries, including the discoverers Cryo-Cell International and MediStem Laboratories, have quickly recognized the great potential of these cells and are not only researching their use in prospective applications, but also offering to cryogenically preserve patients' menstrual blood. For more information on these companies, see their websites at http://www.cryo-cell.com/ and http://www.medisteminc.com/, respectively.

The significant promise these cells hold as a large reservoir of patient-specific stem cells that can be obtained in a relatively easy manner may make people think twice about discarding any cells that were once a part of their body.

Further Reading

The following resources may be of interest for those wanting to learn more about endometrial regenerative cells, mesenchymal stem cells, and stem cells in general:

- Teisha J. Rowland's blog "All Things Stem Cell" at http://www.AllThingsStemCell.com
- Teisha J. Rowland's blog article "Stem Cells Discovered in Menstrual Blood: Endometrial Regenerative Stem Cells" at http://www.allthingsstemcell.com/2009/03/endometrial-regenerative-stem-cells/
- Teisha J. Rowland's blog article "Mesenchymal Stem Cells: A Diverse Family, Large and Still Growing" at http://www.allthingsstemcell.com/2009/03/mesenchymal-stem-cells/
- The National Institutes of Health's webpage "Stem Cell Information: Frequently Asked Questions (FAQs)" at http://stemcells.nih.gov/info/faqs.asp
- Teisha J. Rowland's webpage "Visual Stem Cell Glossary" at http://www.AllThingsStemCell.com/Glossary
- Man-Jing Zhang et al.'s article "Could cells from menstrual blood be a new source for cell-based therapies?" at http://dx.doi.org/10.1016/j.mehy.2008.10.021

Hans Keirstead, a Stem Cell Game-Changer

Involvement in the first FDA-approved clinical trial using human embryonic stem cells

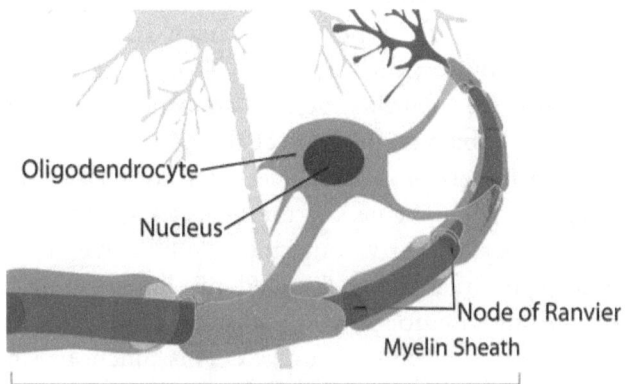

The Axon of a Neuron

Dr. Hans Keirstead helped develop methods to generate oligodendrocytes from human embryonic stem cells. Oligodendrocytes protect a neuron's axon by insulating it in a myelin sheath. When the spinal cord is injured, oligodendrocytes die, leaving the axon unprotected. (In this drawing, the cell body of the axon is out of the picture, to the left.)

DR. HANS KEIRSTEAD

Game-changing discoveries have always been met with resistance. This has been clearly shown by accomplishments achieved by prominent stem cell researcher Dr. Hans Keirstead. Keirstead is an associate professor of Anatomy and Neurobiology and co-director of the Sue and Bill Gross Stem Cell Center at the Reeve-Irvine Research Center at the University of California, Irvine. Keirstead and his lab developed a method using human embryonic stem cells (hESCs) to treat spinal cord injuries and, in 2009, this treatment became the first-ever FDA-approved clinical trial using hESCs.

SPINAL CORD INJURIES AND OLIGODENDROCYTES

During his Ph.D. studies at the University of British Columbia in Vancouver, Canadian-born Keirstead developed ground-breaking methods to regenerate damaged spinal cords. To understand how these methods work, it's important to have an understanding of how the spinal cord normally functions. Two key groups of cells in the spinal cord are neurons, which transmit information, and oligodendrocytes, which act to protect the neurons and ensure that the information travels properly. (The latter are a type of glial cell, so named for being the "glue" of the nervous system). The main way oligodendrocytes carry out their supportive roles is by insulating neurons in a material called myelin, forming a myelin sheath around the neuron's axon, a long protrusion from the neuron that conducts electrical impulses. This functions similarly to how an electrical wire must be insulated to work properly.

However, when the spinal cord is injured, oligodendrocytes die en masse and consequently the neurons lose their protective myelin sheaths. Even many oligodendrocytes far from the site of impact can die, increasing the area of the spinal cord that has neurons without myelin sheaths. The absence of the myelin sheaths so greatly disrupts the flow of information across the neurons in the spinal cord that paralysis can result even when the neurons themselves are spared.

Consequently, the main approach Keirstead and others have been taking to regenerate damaged spinal cords is to restore the myelin sheaths to, or "remyelinate," the neurons. This has been primarily accomplished by introducing new oligodendrocytes to the damaged spinal cord area.

A HUMAN EMBRYONIC STEM CELL-BASED THERAPY

Keirstead continued developing methods of treating spinal cord injuries using oligodendrocytes, but incorporated hESCs into his approaches when he joined the Reeve-Irvine Research Center at UC Irvine. Four years after joining the Center, in 2005, Keirstead reported that his group could not only create oligodendrocyte progenitors from hESCs, but could also use these progenitors to effectively treat rats with acute spinal cord injuries. This was published in *The Journal of Neuroscience*, and to read this article see http://dx.doi.org/10.1523/JNEUROSCI.0311-05.2005. Specifically, when these rats were injected with oligodendrocyte progenitor cells (termed OPCs) made from hESCs, the rats were able to remyelinate neurons and improve motor functions.

While human embryonic stem cells have immense potential for the field of regenerative medicine because they are seemingly unlimited in numbers and are pluripotent (meaning they can become any type of cell), they have also been surrounded by ethical debate due to their biological origins. But, other than stem cells from fetal tissues (fetal mesenchymal stem cells specifically, which are from a much later stage in development than hESCs are) or neural stem cells (which are obviously quite limited in supply), hESCs are the only type of stem cell that has been found to be able to become oligodendrocytes (or oligodendrocyte progenitors). Consequently, hESCs are currently thought to be the best source of these cells for spinal cord therapies. And they appear to be quite effective.

FIRST FDA-APPROVED CLINICAL TRIAL USING hESCS

On January 23, 2009, after submitting very promising data from 24 studies in animals in a 21,000-page application (hopefully it didn't need to be printed!), Keirstead and collaborators at the biopharmaceutical company Geron received FDA approval for the first-ever human trial using hESCs. To learn more about Geron, see http://www.geron.com.

The Phase I clinical trial was to treat patients with subacute thoracic spinal cord injuries by injecting the hESC-derived OPCs (labeled "GRNOPC1") into the site of injury in the patient's spinal cord, as this method showed significantly improved neuron remyelination and locomotion in animals for over nine months after a single injection with OPCs. Seven medical centers in the U.S. were selected for the clinical trial, and eligible patients would have to have had a spinal cord injury between the T3 and T10 thoracic spinal segments (near the middle of the spinal cord) and be able to be treated with OPCs only 7 to 14 days after the injury. The short time window of eligibility for treatment is clearly an aspect requiring improvement, but reflects what was observed in the animal models: quick action is needed after injury for these methods to be effective. Hopefully future approaches can be developed, or current ones improved upon, to increase the time window of treatment.

However, in August of 2009, before any patients had actually been treated, the trial was put on hold by the FDA. Additional animal data submitted by Geron revealed that small, non-proliferative (they do not grow) cysts were found in the injury site, although the presence of these cysts did not correlate with animals having negative side effects. In late October, 2009, Geron announced that in the near future they would reinitiate the Phase I clinical trial in patients with thoracic spinal cord injuries, and might expand to include patients with cervical (the top part of the spinal cord) injuries in the future.

In June 2010, Keirstead gave a presentation in Santa Barbara, California, on "Stem Cell Research and its Application for Treatment of Spinal Cord Injury and Disease" and I was lucky enough to be able to attend. He discussed the biology behind these potential spinal cord injury cures, as I've done here, and the future direction of the therapies. Soon after his talk, in July 2010, Geron announced that the trial would reinitiate, after an additional study characterizing the OPCs in animals had been completed. Geron had also been exploring the applicability of the OPCs in treating other neurological diseases, such as Alzheimer's disease, strokes, and multiple sclerosis, all additional important avenues of research.

However, in November 2011, Geron announced that the clinical trial would be closed to additional enrollment. Not only that, but Geron also stated that they'd withdraw from the stem cell field entirely. The primary reasons given were financial and to focus more on their cancer research, specifically a promising telomerase inhibitor. (Telomerase and cancer is discussed later, in Chapter 4.) When this was announced, four patients had already each received a one-time injection. While it was too early to know whether the treatment had been effective, preliminary results indicated that no adverse effects had been observed. These patients, and others who had already received approval to participate in the trial, would still receive treatment and be monitored. In January 2013, it was announced that Geron's hESC assets, including the FDA-approved Phase I clinical trial, were being bought by BioTime Inc., another California-based company. This purchase should help make BioTime an even more prominent leader in the regenerative medicine field.

THE FUTURE OF hESC-BASED THERAPIES

The location of Keirstead's June 2010 presentation was very relevant for the stem cell community. The University of California at Santa Barbara (UCSB) is becoming a hub for stem cell research that is transitioning to the clinic. Interdisciplinary groups on campus collaborate and use hESCs and other stem cells for regenerative medicine applications in the UCSB Center for Stem Cell Biology and Engineering. As a graduate student in Dr. Dennis Clegg's laboratory at UCSB, I was able to be involved in such interdisciplinary collaborations, extensively use the Center, and watch it expand over the years as more groups came to utilize it. For more on the Center, see http://www.stemcell.ucsb.edu. Additionally, Dr. James Thomson, the researcher who originally isolated hESCs in 1998, holds a joint faculty appointment at the University of Wisconsin and UCSB. Clearly, UCSB is at the forefront of a wave of burgeoning stem cell facilities, and Dr. Keirstead's talk found a receptive and motivated audience.

As knowledge and awareness of the uses of these very promising stem cells has continued to grow, the resistance they originally met with has already been diminishing. Collaborative efforts with biopharmaceutical companies may ultimately be key to demonstrating the true potential of these cells in treating human diseases and injuries. While the relationship between Keirstead's work and his Geron collaborators did not have an ideal ending, it is hoped that BioTime can pick up where Geron left off, successfully complete the initial clinical trial, and ultimately fulfill what Keirstead and Geron set out to do. It is also important not to forget that Kerstead and Geron's relationship was important in another way – by receiving the first FDA-approval of a clinical trial using hESCs, Geron helped pave the way for other researchers seeking FDA-approval for different clinical trials that also use hESCs. This was certainly a significant achievement. Overall, this story shows that, once again, it is only with considerable support that these kinds of stem cell-based treatments can reach the individuals who need them.

Further Reading

The following resources may be of interest for those wanting to learn more about Keirstead's spinal cord injury treatments, human embryonic stem cells, and stem cells in general:

- Teisha J. Rowland's blog "All Things Stem Cell" at http://www.AllThingsStemCell.com
- Teisha J. Rowland's blog article "Human Embryonic Stem Cells" at http://www.AllThingsStemCell.com/2009/04/human_embryonic_stem_cells/
- California Stem Cell, Inc.'s webpage "Hans. S. Keirstead, Ph.D." at http://www.californiastemcell.com/about-us/scientific-advisory-board/hans-s-keirstead-ph-d/
- Hans Keirstead et al.'s article "Human Embryonic Stem Cell-Derived Oligodendrocyte Progenitor Cell Transplants Remyelinate and Restore Locomotion after Spinal Cord Injury" at http://dx.doi.org/10.1523/JNEUROSCI.0311-05.2005
- Geron's press releases at http://ir.geron.com/phoenix.zhtml?c=67323&p=irol-news&nyo=4
- Jef Akst's article "First hESC Trial Kaput" in *The Scientist* magazine at http://www.the-scientist.com/?articles.view/articleNo/31391/title/First-hESC-Trial-Kaput/
- Cynthia Dizikes's article "Stem cell trial's cancellation disappoints paraplegic patient and a Northwestern researcher" in the *Chicago Tribune* at http://articles.chicagotribune.com/2012-01-10/news/ct-met-stem-cell-trial-20120110_1_cell-research-geron-cell-transplantation-program
- Beth Marie Mole's article "Geron Sells Stem Cell Assets" in *The Scientist* magazine at http://www.the-scientist.com//?articles.view/articleNo/33876/title/Geron-Sells-Stem-Cell-Assets/
- The National Institutes of Health's webpage "Stem Cell Information: Frequently Asked Questions (FAQs)" at http://stemcells.nih.gov/info/faqs.asp
- Teisha J. Rowland's webpage "Visual Stem Cell Glossary" at http://www.AllThingsStemCell.com/Glossary

How to Make Your Own Stem Cells

There are now multiple ways adult stem cells can be turned embryonic

Somatic cell nuclear transfer (SCNT) was famously used to create the first cloned animal from an adult cell, Dolly the sheep (pictured above). SCNT can also be used to turn adult cells into embryonic stem cells, and such "therapeutic cloning" techniques hold much clinical promise as sources of patient-specific cells.

There are many reasons a patient may need a tissue or blood transplant from a donor, and in many cases the problem of immune rejection is a major obstacle to be overcome. Consequently, researchers have increasingly focused on developing ways to heal damaged or diseased tissues by using a patient's own cells. There are now multiple types of stem cells that hold great promise for creating patient-specific cells for therapies.

Rejection of a transplant from a donor occurs when a patient's immune system rejects the transplant tissue as foreign. It's a normal (and usually healthy) process. The body is simply trying to protect itself from a foreign "invader," just as it usually fights off viral and bacterial infection. Immune rejection is a common problem, depending on the kind of transplant being performed, and often long-term courses of immunosuppressant drugs are required over the patient's lifetime to prevent rejection of the transplant.

But by using a patient's own cells to repair the tissue, there should be no immune response triggered. At this point in time, two types of stem cells hold great promise for use in regenerative medicine where immune rejection is a significant concern: human induced pluripotent stem cells (iPSCs) and cells made through a process called somatic cell nuclear transfer (SCNT).

Both of these types of cells have the ability to become any type of cell in the body, and both can multiply virtually indefinitely. These two characteristics are what give such promise to human embryonic stem cells (ESCs). But unlike ESCs, iPSCs and cells from SCNT can be made using a patient's own cells, making them patient-specific and avoiding immune rejection complications. Because iPSCs were focused on previously in this chapter, here we will focus on advances in SCNT technology and its therapeutic potential.

SOMATIC CELL NUCLEAR TRANSFER

SCNT technology significantly predates iPSCs, and, as was discussed previously, in many ways SCNT technology formed the basis for the idea of iPSCs. In SCNT, the nucleus (which holds the cell's DNA) is taken from a matured, adult cell (all such cells are called "somatic" cells). This nucleus is implanted into an egg that already had its own nucleus removed (it's what's called an enucleated egg). Amazingly, the egg "reprograms" the nucleus to an embryonic state, together creating a new embryo but with the adult cell's genetic lineage. It's been found that SCNT causes some 10,000 to 12,000 genes to be activated that are normally only active during embryonic development. The newly formed embryo (technically called a blastocyst) can then be implanted into a surrogate mother, and potentially become an adult organism. This organism is a clone of the animal that donated the nucleus; they have identical DNA. This whole process is shown in the diagram below.

Although nuclear transfer studies have been conducted since the late 1930s, it wasn't until 1997 that the first widely-accepted-as-successful use of SCNT was reported: Dolly the sheep was born, and she was the first cloned animal from an adult cell. (Dolly's remains are currently on display in the National Museum of Scotland, in Edinburgh, where I was thrilled to visit her in 2012.) Since Dolly, several other animals have been successfully cloned, including rabbits, endangered wild sheep, and pigs, and mice have even been cloned and re-cloned for 25 generations from the same original source, though problems still remain. (Usually only about zero to 10 percent of embryos created through SCNT result in live births).

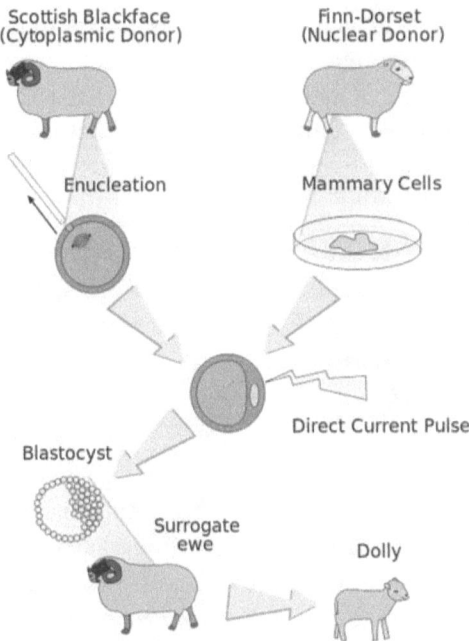

Dolly the sheep was cloned through SCNT. An adult cell from the mammary gland of a Finn-Dorset ewe acted as the nuclear donor; it was fused with an enucleated egg from a Scottish Blackface ewe, which acted as the enucleated egg donor. A successfully implanted oocyte developed into Dolly, a clone of the nuclear donor, the Finn-Dorset ewe.

THERAPEUTIC CLONING

While SCNT has been long-explored for its ability to create cloned animals like Dolly and others (a practice called "reproductive cloning"), SCNT has other very appealing applications that do not involve the creation of an entire animal – primarily "therapeutic cloning." The goal of therapeutic cloning is to use SCNT technology to create patient-specific embryonic stem cells for medical therapies, such as regenerative medicine.

How can SCNT be used to create embryonic stem cells (ESCs)? Normally, ESCs are harvested from an embryo at the blastocyst stage. That's when the embryo is four or five days post-fertilization and is a fluid-filled sphere containing approximately 150 cells. As was described above, SCNT creates embryos that are clones of the DNA donor. When a cloned embryo is at the blastocyst stage, ESCs can be harvested from the blastocyst just as they normally are.

While using SCNT to clone an entire animal has been fraught with challenges, using SCNT to create ESCs for therapies may be less difficult since it only requires the formation of a very early-stage embryo. SCNT-derived ESCs have now been created from multiple types of animals.

SCNT IN THE MOUSE

In 2000, the first ESCs were made from mice using SCNT. These SCNT-derived mouse ESCs behaved similarly to, and had the same abilities as, normally-derived mouse ESCs. Not only could ESCs be successfully derived in this way, but studies in mice have also shown that SCNT can be used to repair genetic diseases, such as Parkinson's or deficient immune systems.

Generally what is done in these studies is this: A mouse with a genetic defect undergoes SCNT. That is, it has the nucleus from one of its adult cells removed and placed into an enucleated mouse egg cell, ultimately creating ESCs. In these ESCs, the defect can be fixed using standard genetic techniques, and the repaired cells (after they have been made into the proper cell type) are transplanted back into the original donor mouse. Mice have been successfully cured of genetic diseases this way.

SCNT IN NON-HUMAN PRIMATES

In 2007, ESCs were created from a rhesus macaque, an Old World monkey, using SCNT. To accomplish this, Shoukhrat Mitalipov and colleagues at Oregon National Primate Research Center took the nucleus from a skin cell of a female rhesus macaque and had the nucleus undergo SCNT in the eggs of another female macaque. Although the isolated ESCs looked and acted like normal ESCs, the process was far from perfect; the researchers had to use 304 eggs to make two lines of ESCs (this means they successfully isolated ESCs from two different blastocysts).

However, progress was quickly made and two years later Mitalipov's group reported a three-fold increase in ESC generation from rhesus macaques using improved SCNT techniques. These levels are similar to those seen with the generation of ESC lines from normally fertilized embryos, giving researchers hope for transitioning the technology to humans.

HUMANS AND SCNT

Currently, a scarcity of high-quality donated human eggs is the largest hurdle to creating patient-specific human ESCs through SCNT. There are also many technical issues that need to be optimized. But one of the greatest challenges is not technical, but social: In 2004, Woo Suk Hwang and colleagues in South Korea reported having created human ESCs through SCNT for the first time, but their work was proven to be fraudulent. Because of these falsified studies, subsequent researchers have had to go to great lengths to show that their SCNT data is valid, making it an even more difficult field in which to achieve progress.

Because of the lack of donated human eggs, researchers have pursued studies using eggs donated from other animals for SCNT. For example, one such SCNT study used human nuclei and rabbit eggs. This kind of SCNT technique is called interspecies SCNT (iSCNT). Many ethical concerns have been raised over the use of this technique, especially since a human-animal iSCNT embryo contains a human nucleus but also some other animal DNA (from the mitochondria). Additionally, some reports have suggested that iSCNT embryos are not "healthy," primarily because the nucleus doesn't completely return to an embryonic state. In one such study, a group of researchers combined the nucleus of an adult chimpanzee cell with an egg from a cow, but the embryos did not grow past having only 16 cells. It turned out that many genes that are essential for early development weren't actually active. Overall, there are still many ethical and biological concerns surrounding the practice of iSCNT.

Despite the lack of available, donated human eggs and lingering pall on the field from the fraudulent work of Hwang and colleagues, in 2008 several stem cell-oriented companies (including Stemagen Corporation and The Reproductive Sciences Center, both located in California) collaboratively published the successful creation of human blastocysts from SCNT. In this study, 21 human eggs were donated with informed consent and each was combined with a nucleus from an adult skin cell through SCNT. From these 21 eggs and nuclei, five blastocysts were created, and at least one was proven to be a true clone of the donor.

THE FUTURE OF SCNT

Although much work remains to be done before SCNT is a viable strategy for the production of patient-specific ESCs for regenerative medicine, there is enormous potential that such "therapeutic cloning" could fulfill when the technology matures. Not only may patients have cells made that are specific to them and available as virtually any cell type they should need (such as insulin-producing beta cells for a patient with diabetes), but, for example, if they have a genetic disease this too may be cured (as was demonstrated in mice using this technology).

Researchers could also create cell lines from patients to study specific diseases in the laboratory, greatly advancing our understanding of these conditions (this is currently being explored with iPSCs as well, as was discussed earlier in this chapter).

Lastly, as our ability to create ESCs from embryos improves in general, it should become easier to create SCNT-derived ESCs. As SCNT research has come close to being completely illegal in the United States, it's important not only for researchers to always take proper ethical precautions, but for everyone to be aware of the great potential this technology holds, and to understand the big difference between "reproductive cloning" (e.g., the creation of Dolly the sheep) and "therapeutic cloning."

Aside from stem cells derived from SCNT, there is a great variety of other stem cell types and technologies that hold much promise in the field of regenerative medicine, as has been discussed in this chapter already. In particular: the very closely-related induced pluripotent stem cells; human embryonic stem cells, which were recently used in initial clinical trials for treating spinal cord injuries; an assortment of adult stem cells; and the recently developed technique of "direct reprogramming."

Further Reading

The following resources may be of interest for those wanting to learn more about somatic cell nuclear transfer and stem cells in general:

- Teisha J. Rowland's blog "All Things Stem Cell" at http://www.AllThingsStemCell.com
- Teisha J. Rowland's blog article "Creating Patient-Specific Stem Cells through Somatic Cell Nuclear Transfer" at http://www.AllThingsStemCell.com/2010/11/creating-patient-specific-stem-cells-through-somatic-cell-nuclear-transfer
- Sayaka Wakayama et al.'s article "Successful Serial Recloning in the Mouse over Multiple Generations" at http://dx.doi.org/10.1016/j.stem.2013.01.005
- Congresswoman Diana DeGette's book *Sex, Science, and Stem Cells*
- The National Institutes of Health's webpage "Stem Cell Information: Frequently Asked Questions (FAQs)" at http://stemcells.nih.gov/info/faqs.asp
- The Human Genome Project's webpage "Cloning Fact Sheet: Cloned Animals" at http://www.ornl.gov/sci/techresources/Human_Genome/elsi/cloning.shtml#animals
- Teisha J. Rowland's webpage "Visual Stem Cell Glossary" at http://www.AllThingsStemCell.com/Glossary

Baldness and the Potential of Progenitors

Hair progenitor cells may hold the keys to treating baldness

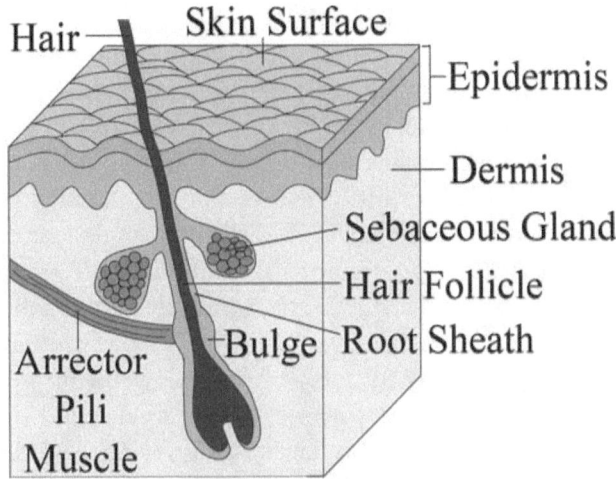

Every hair sits inside a hair follicle, which goes down through the epidermis and dermis of the skin. The sebaceous gland releases oils and the arrector pili muscle can cause a hair to stand on end. The bulge is where the majority of the hair follicle's stem cells reside, and these can give rise to progenitor cells that travel to the bottom of the hair follicle when a new hair starts to grow, giving rise to the cells needed to make the new hair.

It's known that stem cells, the key players in regenerative processes in the body, play a major role in continually making new hair. Naturally, this role caused researchers to suspect that hair stem cells are key players in causing androgenetic alopecia (AGA), or male pattern baldness, the most frequent type of hair loss among men. (For more information on male pattern baldness, see http://www.ncbi.nlm.nih.gov/pubmedhealth/PMH0002160.)

However, in 2011 a group of researchers showed, surprisingly, that patients with AGA actually had a normal amount of hair stem cells in their scalps. Instead, the team found that different "progenitor cells" may actually be the cells whose fates are tied to AGA. This better understanding of the exact cell types involved may ultimately help researchers devise better therapies for treating male pattern baldness.

ANDROGENETIC ALOPECIA

In order to understand how stem cells or "progenitor cells" may be involved in AGA, it's important to know how hair usually grows, and how this process is changed in AGA. At any given time, about 85 percent of the hair follicles on a person's head are in a growing phase. (Hair follicles are little cavities that go down through the dermis skin layer, and each cavity is home to a single hair, as shown in the diagram at the beginning of this section.) This growth phase can last two to six years for any given hair, during which time the hair grows about five inches every year. (Of course, there's some variation from person to person.) After this growing period, the hair follicle and root shrink for one to two weeks, and during the following five to six weeks the hair stops growing. At the end of this whole process, the hair follicle re-enters the growing phase, sometimes continuing to lengthen the original hair and sometimes with a newly growing hair pushing the old one out, starting the growth cycle all over again.

Normally, the new hair will grow similarly to the previous hair. However, with AGA this isn't the case. In AGA, hair follicles get smaller over time, and consequently make smaller and smaller, eventually microscopic, hairs. How is this baldness caused? It's not that well understood; it's known that testosterone is necessary for this miniaturization, but not much else is known about what causes AGA.

BULGE STEM CELLS

While the ultimate cause of AGA remains, thus far, elusive, researchers have done a great deal of work to determine the presence and role of stem cells in normal hair follicles. Their work has lead to the present understanding that, in a hair follicle, there are stem cells that reside in a small compartment called the "bulge." (The bulge is labeled in the diagram at the beginning of this section.) These stem cells can turn into nearly all of the different cell types found in the hair follicle. They're intimately involved in

70

the hair follicle life-cycle; if these stem cells are destroyed, so is the hair follicle.

A study by George Cotsarelis and colleagues at the University of Pennsylvania School of Medicine, published in 2011 in *The Journal of Clinical Investigation*, set out to resolve whether these bulge stem cells play a key role in AGA. (To read the published study, see http://dx.doi.org/10.1172/JCI44478.) To do this, the team took samples of skin cells from the heads of people with AGA, and compared samples from areas that were bald to areas that were haired (non-bald). What did they "compare" exactly? To figure out what kinds of cells were present in their samples, the researchers looked at what proteins were in the samples. Certain cells make certain proteins. For example, researchers can identify if their sample has stem cells from the bulge by looking for the presence of certain proteins known to be made only by those cells. After all of this preparation, the researchers found that, surprisingly, the percentage of bulge stem cells was the same in both the bald and haired samples. The presence of the bulge stem cells isn't enough to prevent hair follicle miniaturization in AGA.

PROGENITOR HAIR CELLS

With their main suspect, the bulge stem cells, released from suspicion due to the new evidence, the researchers realized that other cellular players may be the real culprits behind male pattern baldness. Their suspicions quickly shifted to some "progenitor cells." Progenitor cells are like stem cells in that they can turn into several different types of cells, but, unlike stem cells, progenitors can divide and multiply a relatively limited number of times (stem cells can replicate many times); progenitors have a much more limited lifespan.

In the hair follicle, the bulge stem cells are known to give rise to different groups of hair progenitor cells. Usually the progenitors travel to the bottom of the hair follicle when a new hair starts to grow, giving rise to the cells necessary for the new hair and acting as a kind of "middle man" between the stem cells and the final desired cell types.

To determine whether progenitor cells were connected with AGA, the researchers started looking for the presence of other proteins in their bald and haired cell samples, proteins known to be made by different groups of cells in hair follicles. Sure enough, they found a connection; they discovered a group of cells that was present in the haired samples but absent in the bald samples. Based on the activity level and size of these cells, the researchers think this population of cells could indeed be a group of progenitors.

Furthermore, by looking at what other proteins these progenitors made, the researchers were able to narrow down their location within the follicle: They live in the bulge, as well as an area near the bulge known to house progenitors. Altogether, it's very likely that these potential culprits are progenitors derived from the bulge stem cells. Importantly, the team also showed that these progenitors could produce all of the cell lineages found in a hair follicle, an indication that their absence could certainly lead to problems with hair growth. Overall, this work cataloged a group of progenitor cells that had not been previously identified and found that these novel cells likely have a key role in the onset of AGA, male pattern baldness.

THE POTENTIAL OF PROGENITORS

These recent findings reported by Cotsarelis and colleagues not only help us better understand how baldness in AGA may be caused, but also how AGA may be treated. The authors hypothesize that in AGA, bulge stem cells that have somehow become inactivated may result in a loss of the essential progenitor cell populations. Because the novel progenitors were shown in their report to be able to generate a number of important hair follicle cell lineages, such a loss of their numbers may lead to the miniaturization of the hair follicle, a hallmark sign of AGA.

But because there are still bulge stem cells, which can make new progenitor cells, present in all of the hair follicles of patients with AGA, it suggests that the AGA condition may be reversible. By better understanding these complex and varied progenitor cell populations, researchers may be able to determine what signals are necessary to coax the remaining bulge stem cells to become the necessary progenitor cells for hair growth, and consequently a promising cellular treatment for AGA may be waiting just over the horizon.

Further Reading

The following resources may be of interest for those wanting to learn more about androgenetic alopecia, hair follicle stem cells and progenitors, and stem cells in general:

- Teisha J. Rowland's blog "All Things Stem Cell" at http://www.AllThingsStemCell.com
- Teisha J. Rowland's blog article "Progenitor Hair Populations are Key to Understanding Male Pattern Baldness" at http://www.AllThingsStemCell.com/2011/02/hair-progenitors-and-baldnes
- Luis A. Garza et al.'s article "Bald scalp in men with androgenetic alopecia retains hair follicle stem cells but lacks CD200-rich CD34-positive hair follicle progenitor cells" at http://dx.doi.org/10.1172/JCI44478
- The U.S. National Library of Medicine's article "Male pattern baldness" at http://www.ncbi.nlm.nih.gov/pubmedhealth/PMH0002160/
- Erin M. Campbell's article "Hair follicle stem cells – the hairy truth" in The Company of Biologists' blog "The Node" at http://thenode.biologists.com/hair-follicle-stem-cells/
- MedicalExpress's article "Sensory hair cells regenerated, hearing restored in mammal ear" at http://medicalxpress.com/news/2013-01-sensory-hair-cells-regenerated-mammal.html
- The National Institutes of Health's webpage "Stem Cell Information: Frequently Asked Questions (FAQs)" at http://stemcells.nih.gov/info/faqs.asp
- Teisha J. Rowland's webpage "Visual Stem Cell Glossary" at http://www.AllThingsStemCell.com/Glossary

CHAPTER 2

How To Make A New Body

Engineering Organs

Whole human airways have been bioengineered using stem cells

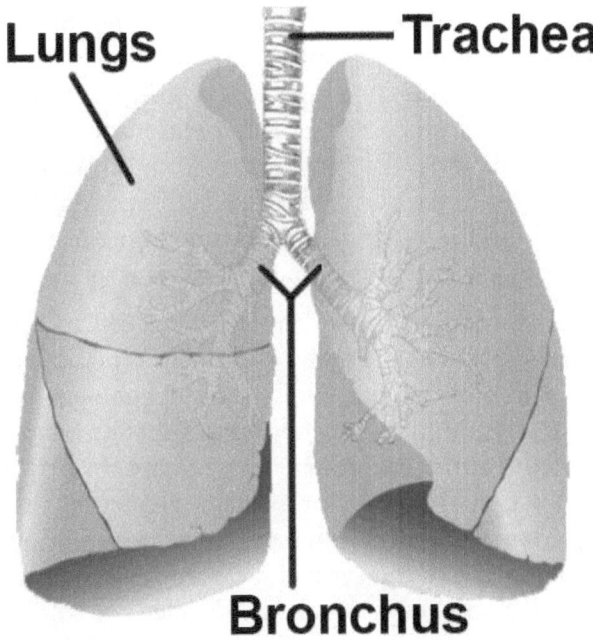

A human airway (specifically a bronchus) has been successfully "bioengineered" using a patient's own stem cells. This was the first human organ made using stem cells.

Demand almost always surpasses supply, often greatly, when it comes to human organs for patients in need of transplants. This has motivated many researchers to work toward engineering biological tissues and organs (a field called "bioengineering") suitable for transplants. Stem cells, displaying their wide medical utility, have played a significant role is this quickly expanding field.

In 2008, a significant breakthrough occurred when researchers created the first human organ manufactured using stem cells, and then transplanted it successfully into a patient, 30-year-old Claudia Castillo. Due to a severe tuberculosis infection, Castillo's left bronchus (which is where the windpipe has branched off to the separate lungs) had become almost entirely collapsed; breathing was so impaired that she could no longer carry out simple domestic chores.

Castillo's relatively unique and tragic situation led Dr. Paolo Macchiarini (University of Barcelona, Spain) and his colleagues, after exhausting traditional approaches, to test a novel organ transplant therapy on her, which had previously been successful in mice and pigs. As a graduate student, I was fortunate enough to attend the International Society for Stem Cell Research 2009 Annual Meeting in Barcelona, Spain, where I saw Macchiarini give an engaging talk on this cutting-edge treatment, the results of which are very promising to the burgeoning field of regenerative medicine.

To treat Claudia, Macchiarini and his team needed to remove and replace her failed bronchus entirely. Normally, replacement of large airway pieces and other organs is a significant problem because the patient must remain on immunosuppressant medications for life to prevent rejection of the new donated tissue, and this can shorten a patient's lifespan by 10 years or more. However, the risks of immune rejection were overcome by, amazingly, using Castillo's own stem cells to create the synthetic organ. For a visual diagram of the treatment process, which is detailed in the next few paragraphs, see

http://www.allthingsstemcell.com/2009/10/bioengineering-organs-breakthroughs/.

THE DONATED AIRWAY

To create the replacement bronchus, a donated airway from a cadaver was obtained and treated so that all donor's cells were removed. The specific piece of donated airway was a segment of trachea. After all connected tissues were removed from the it, the trachea was extensively treated to remove any cells (to be used in a transplant, all "foreign," or donor, cells must be removed to prevent immune rejection by the patient). This procedure took over 6 weeks, and involved 25 cycles of the trachea being incubated with detergents and enzymes that break down DNA. After this laborious process, only the cartilage of the trachea remained, leaving a hollow, bleached tube.

CASTILLO'S OWN STEM CELLS

This blanched trachea, now devoid of any donor tissues and cells, acted as the perfect scaffold for Castillo's own cells to be grown on. Two different kinds of cells were taken from Castillo and grown on the trachea: epithelial cells (cells that cover the inside and outside of organs) and chondrocytes (cartilage cells). The epithelial cells were taken from the moist tissue lining (mucosa) of Castillo's healthy bronchus and were coaxed to grow on the inside of the donor trachea. Obtaining the chondrocytes was a bit trickier; the researchers created the chondrocytes from a population of stem cells in Castillo's body. From a sample of Castillo's bone marrow, the researchers isolated out mesenchymal stem cells, which were discussed in Chapter 1, and then turned these stem cells into chondrocytes in just three days (using an established procedure of exposing the cells to specific molecular factors). These chondrocytes were grown all over the outside of the trachea.

The epithelial cells, on the inside of the trachea, and the chondrocytes, on the outside of the trachea, were both grown in their own different environments (exposed to different growth factors and conditions) that mimicked what they would encounter in the human body. This was done in a "bioreactor," a box where the inside and outside of the trachea were exposed only to their two separate environments. The cells were grown on the trachea in the bioreactor for four days, at which point the researchers had created, or bioengineered, a human airway lacking any synthetic parts and populated by Castillo's own cells.

A SUCCESSFUL TRANSPLANT

Once the bioengineered trachea was ready, the near-collapsed portion of Castillo's left bronchus was removed and replaced by the trachea, now acting as a segment of bronchus. A month after the transplant, the trachea was indistinguishable from Castillo's normal right bronchus and the surrounding bronchus tissue. The transplanted airway displayed completely normal function. In 2009, a year later, it was reported that the graft and patient are still doing fine. Castillo is not only able to do daily chores, but even to dance!

SUCCESS AGAIN: MANIPULATING STEM CELLS AFTER A TRANSPLANT

In March of 2010, this same strategy was successfully used in a 10-year-old boy. Dr. Macchiarini, who is an honorary professor at the University College London (UCL), worked with doctors and researchers at UCL, the Great Ormond Street Hospital for Children (GOSH) in London, the Careggi University Hospital in Florence, and others, to perform the transplant. The boy had a rare condition called long segment congenital tracheal stenosis, which resulted in the boy having an extremely narrow trachea that does not grow. As with Castillo, the doctors used stem cells from the boy's bone marrow and grew them on a cadaveric trachea (that had been stripped of its cells). However, unlike the transplant done with Castillo, with the boy the doctors added factors to prompt the stem cells to mature after they were transplanted into the boy (instead of maturing them entirely in a bioreactor), making the process much faster.

THE FUTURE OF BIOENGINEERING

Beyond the work of Macchiarini and colleagues, other research groups have made breakthroughs in bioengineering different organs and tissues in the past few years. Researchers have been able to create and shape muscle segments using silicon-based synthetic scaffolds in the laboratory. Others have discussed the potential of using stem cells to rescue damaged heart muscles or create bioengineered intestines for transplantation. Mesenchymal stem cells (as used in the above airway transplants) hold great potential for healing wounds in general: These stem cells can turn into many different kinds of cells, be collected in large numbers, potentially migrate to areas they are needed in for healing, and be immunosuppressive. The combined use of stem cells with nanomaterials (which can mimic the surface of cells and tissues) holds much potential for future scaffold designs in regenerative medicine.

While successfully bioengineered and transplanted airways are breakthroughs, especially through the use of a patient's own stem cells to prevent immune rejection, improvements will be needed to make such transplants feasible for a greater number of patients.

Because Macchiarini's groups used parts of organs from donors (the original cadaveric trachea segment), the transplants are still limited by available donors. It is hoped that research efforts will lead to fully-tissue-engineered organ transplants without the need of such donor grafts, helping address the current shortage of donor tissues and organs, to more effectively treat a large aging population.

The transition to the clinic of other stem cell-based regenerative therapies will also require extremely careful characterization of each individual procedure. There are still many obstacles to overcome before such therapies can become common practice, and those interested in receiving stem cell therapies should be aware of the possible risks involved; for a discussion of potential hazards, see the U.S. Food and Drug Administration's website on Cellular & Gene Therapy Products at

http://www.fda.gov/BiologicsBloodVaccines/CellularGeneTherapyProd ucts/default.htm or the International Society for Stem Cell Research's websites on Frequently Asked Questions at http://www.isscr.org/home/resources/learn-about-stem-cells/stem-cell-faq or Guidelines for the Clinical Translation of Stem Cells at http://www.isscr.org/home/publications/ClinTransGuide.

Further Reading

The following resources may be of interest for those wanting to learn more about bioengineering organs and stem cells in general:

- Teisha J. Rowland's blog "All Things Stem Cell" at http://www.AllThingsStemCell.com
- Teisha J. Rowland's blog article "Bioengineering Organs and Tissues with Stem Cells: Recent Breakthroughs" at http://www.allthingsstemcell.com/2009/10/bioengineering-organs-breakthroughs/
- S. Baiquera, M. A. Birchall, and Paolo Macchiarini's article "Tissue-engineering tracheal transplantation" at http://dx.doi.org/10.1097/TP.0b013e3181cd4ad3
- The BBC's article "Windpipe transplant success in UK child" at http://news.bbc.co.uk/2/hi/health/8576493.stm
- David T. Harris's article "Opinion: Younger is Better" at http://www.the-

scientist.com/?articles.view/articleNo/32483/title/Opinion--Younger-Is-Better/

- The National Institutes of Health's webpage "Stem Cell Information: Frequently Asked Questions (FAQs)" at http://stemcells.nih.gov/info/faqs.asp
- Teisha J. Rowland's webpage "Visual Stem Cell Glossary" at http://www.AllThingsStemCell.com/Glossary

Re-Growing Limbs

Learning from the salamander

The salamander (*Ambystoma mexicanum*) has the amazing ability to re-grow entire severed limbs. By better understanding the stem cells involved, researchers may help improve wound healing in people.

One of the dreams of medicine is to be able to replace a damaged limb with a healthy one, or, more generally, to improve the body's ability to heal wounds. While humans cannot grow new limbs if one is lost, this amazing ability is found in some other animals, and it is hoped that by better understanding limb regeneration in such animals we will learn how to better heal ourselves.

THE SALAMANDER DEBATE

This is why the salamander (*Ambystoma mexicanum*), a.k.a the axolotl, an animal that can completely re-grow a severed limb that is indistinguishable from the original, is widely studied. However, despite a large body of research over the years, how exactly the salamander accomplishes this remarkable feat has been unclear, and hotly debated.

That was, until recently. Research done by Elly Tanaka and her colleagues (at the Max Planck Institute of Molecular Cell Biology and Genetics and the Center for Regenerative Therapies at the University of Technology, both in Dresden, Germany) has revealed that the mechanism by which salamanders re-grow lost legs is actually quite different from what it was long believed to be.

When a salamander has a limb severed, the resultant stub (called a "limb bud") undergoes a well-defined process to re-grow the lost appendage. (For a visual representation of this process, see http://www.allthingsstemcell.com/2009/08/limb-regeneration/.) First on the scene is the epidermis (the outer layer of skin), which grows and expands to cover the wound, closing it within 24 hours after the injury. This layer thickens and then cells from the immune system (macrophages and neutrophils) arrive to clean up the wound, breaking down the injured tissues, cells, and proteins underneath the epidermis.

MYSTERIOUS STEM CELLS

It is at this point in the regeneration process that stem cells take center stage. As the limb bud becomes home to an increasing number of stem cells, it is re-named the "blastema." For a long time scientists knew that there are stem cells in the blastema, but the exact type of stem cells, where they come from, and how exactly they contribute to regenerating the limb in the blastema were fairly mysterious.

Stem cells are special cells because they can not only renew themselves continually (always creating another stem cell to replace the previous one), but can also turn into other, more mature types of cells. Since the mid-1990s, researchers thought that the blastema contained a homogenous (where all cells are identical) group of pluripotent stem cells. ("Pluripotent" means that each individual stem cell would be capable of becoming any type of cell.) The theory went like this: Mature cells in the tissue surrounding the blastema were (somehow) stimulated by the epidermis and nearby cells to become pluripotent stem cells in order to finish regenerating the limb.

A MIXED POPULATION OF LIMITED STEM CELLS TO THE RESCUE

This model was challenged by experiments done by Elly Tanaka's group, published in 2009. When I attended the International Society for Stem Cell Research 2009 Annual Meeting, I saw Tanaka present these controversial findings, and I was fortunate enough to see her give additional related talks at the University of California, Santa Barbara, since then. In 2009, Tanaka and colleagues revealed that the limb regeneration process in salamanders is actually accomplished by a heterogeneous (mixed) population of multiple different types of stem cells that can actually only become one or two different types of mature cell types (making these stem cells unipotent or multipotent at best, but far from pluripotent).

In their efforts to resolve the long-standing debate, Tanaka and her researchers developed a way to track where each different tissue type was during limb regeneration. What they found was that the blastema actually contains different precursors for all the different tissue types that are found in the complete limb. Specifically, the blastema contained precursor cells for all the mature tissue types, but each precursor could only become one or two specific adult tissue types. So it appears that, rather than the mass of magically-coordinated pluripotent cells once believed to be present, several different types of rather limited stem cells may work together to regenerate the entire limb.

In fact, only one of the tissue types that Tanaka's group looked at was capable of becoming more than one type of mature tissue, and this was the dermis, an inner skin layer. The dermis progenitors could turn into either dermis or cartilage (but not other tissue types investigated). In developing salamander embryos, dermis and cartilage actually share a common tissue origin (specifically called the mesoderm), which helps explain why the dermis-layer cells could become either dermis or cartilage.

LIMB REGENERATION IN OTHER ANIMALS

Similar to salamanders, it's been found that some frogs (of the genus *Xenopus*) also re-grow their limbs using a blastema that has a heterogenous mix of progenitor stem cells. It's still unclear whether other animals, on the other hand, may regenerate themselves by reverting mature cells to a pluripotent state (as was long thought to be done by salamanders), or perhaps by changing one mature cell directly into another mature type.

While a universal regeneration mechanism between different animals known to be able to regenerate their limbs or parts of their limbs remains unclear, what is clear is that such research could be quite useful for future regenerative medicine research in humans. The salamander studies indicate that we may not need pluripotent cells for complex tissue regeneration, but instead could use multiple different stem cells with much more limited potentials. As a variety of different, limited stem cells can be fairly easily harvested from an adult human patient (such as hematopoietic and mesenchymal stem cells, which were discussed in Chapter 1 and in the previous section), it presents a vision of human tissue regeneration. While this does not necessarily make the process immediately feasible, it may help give researchers a more focused direction for future studies.

Further Reading

The following resources may be of interest for those wanting to learn more about limb regeneration in the salamander and stem cells in general:

- Teisha J. Rowland's blog "All Things Stem Cell" at http://www.AllThingsStemCell.com
- Teisha J. Rowland's blog article "Limb Regeneration May Require Less Potent Stem Cells Than Previously Thought" at http://www.allthingsstemcell.com/2009/08/limb-regeneration/
- Elly Tanaka et al.'s article "Cells keep a memory of their tissue of origin during axolotl limb regeneration" at http://dx.doi.org/10.1038/nature08152
- The National Institutes of Health's webpage "Stem Cell Information: Frequently Asked Questions (FAQs)" at http://stemcells.nih.gov/info/faqs.asp
- Teisha J. Rowland's webpage "Visual Stem Cell Glossary" at http://www.AllThingsStemCell.com/Glossary

The Immortal Jellyfish

Turritopsis can change the direction of its life cycle

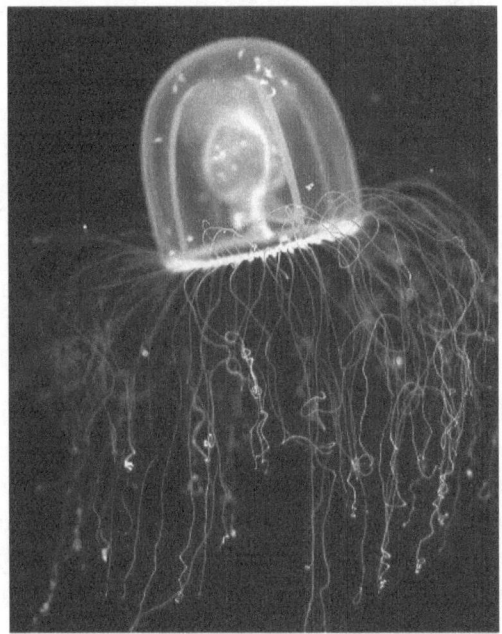

The tiny *Turritopsis nutricula* may have learned the secret of immortality, and, with this innate knowledge, has been leading a silent invasion into oceans around the world.

To be forever young. That has been the elusive dream for people since the dawn of humanity, sparking such myths as the "Fountain of Youth." Yet for one tiny jellyfish, that "dream" is merely the way of life.

TURRITOPSIS

Completely grown-up, adult jellyfish belonging to the genus *Turritopsis* possess an amazing ability: they can reverse their life cycle and turn back into their "newborn" form. This genus belongs to a group called the hydrozoans, which includes some small, predatory jellyfish, hydra, and the Portuguese Man o' War, among others.

While most hydrozoans die after reproducing, *Turritopsis* has other plans. In fact, they're the only animals in the world that have been found to be able to reverse their life cycle like this.

To truly appreciate this remarkable feat, it's important to understand the basic, "normal" life cycle of the *Turritopsis* jellyfish. *Turritopsis'* life cycle is made up of two basic stages: an immature polyp and a sexually mature medusa (also commonly referred to as the jellyfish). When the *Turritopsis* jellyfish create a fertilized egg, it settles to the ocean floor, in shallow or deep waters. The eggs can hatch within a few days. A single egg turns into a polyp, a cup-shaped, sessile (rooted in place), immature form that has tentacles on its raised mouth to help it catch food, and a large central cavity to then digest its food. Many polyps cluster together to form interconnected polyp colonies.

So how do the polyps turn into the adult medusae? Fascinatingly, individual polyps actually form a bud that breaks off and turns into an adult medusa. This is done asexually, without any of the polyps exchanging genetic material. The medusa looks like the classic jellyfish form we're so familiar with: solitary, somewhat umbrella-shaped, and adorned with many tentacles.

TURNING BACK TIME

That's the normal, tried-and-true life cycle for many jellyfish, but *Turritopsis nutricula* and a related species, *Turritopsis dohrnii*, can do this and much more. *T. nutricula* is quite small, about only one-fifth of an inch in diameter, with transparent walls and a vibrantly red-colored large stomach. It can have an astonishing amount of tentacles, around 80 to 90. But its appearance isn't what's gained it so much attention; it's what it does. Under certain conditions, *T. nutricula* can successful transform to its juvenile stage 100 percent of the times tested in laboratories.

What does this mean in terms of the seemingly "bizarre" jellyfish lifestyle? The sexually mature, adult medusa can actually transform into a ball of tissue, like a cyst. It does this by somehow reabsorbing the external parts of its body. The cyst attaches to the ocean floor and, within a few days, depending on the temperature, grows into a colony of immature cylindrical polyps, and the cycle starts all over again; new medusae bud off of the polyps, forming genetically identical jellyfish. The cysts can even wait for months before turning into polyps, if temperatures are very cool. Consequently, they're virtually immortal. But this reversion doesn't happen all the time: It only happens when the jellyfish are starved, physically injured, or under other stresses, and most likely many jellyfish die before undergoing this transformation.

TRANSDIFFERENTIATION

So how exactly does the *Turritopsis* get rid of its wrinkles and gray hairs? It's still a big mystery, but current thought suggests that the process of transdifferentiation is involved. Transdifferentiation is when an adult, mature cell turns into a completely different kind of cell. In the *Turritopsis nutricula* medusa, it's thought, certain adult tissues layers transform into different, specific tissue layers found in the new cyst. While this is somewhat understood on a general tissue level, on the cellular level there is much that remains to be elucidated. Additionally, there is some debate over whether stem cells are necessary for the process of transdifferentiation, and how it works in *Turritopsis* is even more unclear.

Transdifferentiation occurs in other animals as well, though usually only in parts of the animal that are re-growing; *Turritopsis* is probably the only known case of an entire organism undergoing transdifferentiation (though there are other contenders). A long-standing classic example of transdifferentiation is the salamander: Cut off a leg and it re-grows a new one, presumably from the mature cells on the leg stump. However, this was recently challenged by research done by Elly Tanaka (at the Max Planck Institute of Molecular Cell Biology and Genetics and the Center for Regenerative Therapies at the University of Technology, both in Dresden, Germany); it now appears that salamanders use several different kinds of limited stem cells (not mature cells) to regenerate their severed limbs, as was discussed in the previous section. Clearly this is a rapidly evolving field as our understanding of these amazing abilities grows.

As we learn more about how animals such as *Turritopsis* can revert all of their cells to a much earlier stage in life, it may help us better understand human diseases and just the basic definition of what it is to be a cell. Transdifferentiation occurs in humans, but mostly in diseases such as cancer. But this is just one of the many hints, apparent by the existence of *Turritopsis*, that the cells in our adult bodies are not necessarily fixed in their ways. Given the right conditions, they can change their very identities.

And the tiny *Turritopsis* is greatly profiting from its innate knowledge. *Turritopsis* have actually been leading a "silent invasion" of the world's oceans. Research done by Dr. Maria Pia Miglietta (Pennsylvania State University) revealed that, just in recent times, the animals' numbers have been significantly increasing, as her article discusses at http://dx.doi.org/10.1007/s10530-008-9296-0. They have also been rapidly spreading around the world, most likely from water released from the ballasts of ships in ports. While they may only be one-fifth of an inch in diameter and lack a central nervous system, these jellyfish sure seem to have a lot of things figured out.

Further Reading

The following resources may be of interest for those wanting to learn more about transdifferentiation, *Turritopsis'* ability to change from an adult to a juvenile, and stem cells in general:

- Teisha J. Rowland's blog "All Things Stem Cell" at http://www.AllThingsStemCell.com
- *Discover magazine*'s blog post "The Curious Case of the Immortal Jellyfish" by Nina Bai at http://blogs.discovermagazine.com/discoblog/2009/01/29/the-curious-case-of-the-immortal-jellyfish/
- Stefano Piraino et al.'s article "Reversing the Life Cycle: Medusae Transforming into Polyps and Cell Transdifferentiation in *Turritopsis nutricula* (Cnidaria, Hydrozoa)" at http://www.jstor.org/pss/1543022
- Maria Pia Miglietta and Harilaos A. Lessios's article "A Silent Invasion" at http://dx.doi.org/10.1007/s10530-008-9296-0
- Shifaan Thowfeequ, Emily-Jane Myatt, and David Tosh's article "Transdifferentiation in Developmental Biology, Disease, and in Therapy" at http://dx.doi.org/10.1002/dvdy.21336
- Hongbao Ma and Yan Yang's article "*Turritopsis nutricula*" at http://www.sciencepub.net/nature/ns0802/03_1279_hongbao_turritopsis_ns0802_15_20.pdf

- The National Institutes of Health's webpage "Stem Cell Information: Frequently Asked Questions (FAQs)" at http://stemcells.nih.gov/info/faqs.asp
- Teisha J. Rowland's webpage "Visual Stem Cell Glossary" at http://www.AllThingsStemCell.com/Glossary

CHAPTER 3

DEVELOPMENTS IN THE LAB

History of the Cell, Part I

A collaborative effort between innovative people, ideas, and tools and led to its discovery

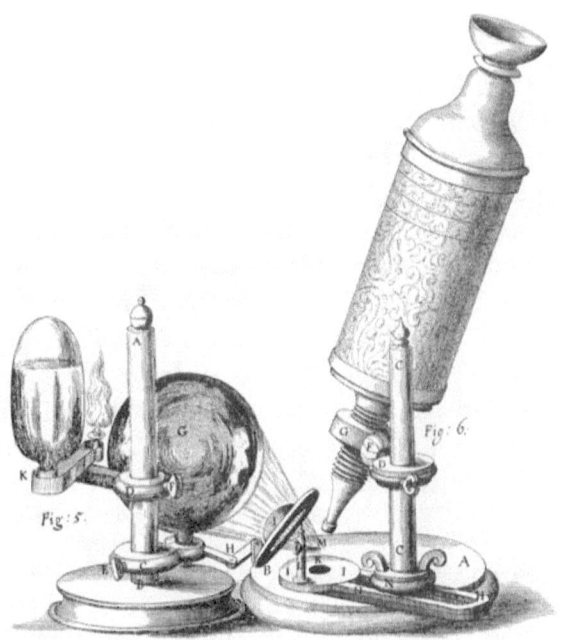

Many great advances in cell biology followed improvements in their tools; in 1665, Robert Hooke first discovered cells with the help of the compound microscope he developed himself (diagrammed by him above).

It's easy to take for granted the amazing scientific discoveries of the last few centuries, or even just the last decade, that were once incredible breakthroughs. One of the most fundamental discoveries for modern biology was the discovery of the cell. From the largest cells – nerve cells as long as 39 feet in giant squids, and the single-celled ostrich egg weighing in at three pounds – to the smallest bacterial cells, all cells share common characteristics and all are the basic building blocks of all life as we know it.

So how did we learn that life is made of cells? Like many scientific discoveries, the first cell observation was not simply due to a few people being more perceptive than their predecessors; it was enabled by many factors, technological and social, falling into place.

THE FIRST MICROSCOPES

Truly, the invention of the microscope, combined with the efforts of many very observant biologists, seeded the field of cell biology. Although obscure, the origins of the microscope most likely lay with the hands of several Dutch lens-grinders in the early 1600s. Scientists throughout Europe quickly snatched up this amazing new tool and carefully recorded and shared their many observations of the world around them that had previously been inaccessible. The first microscopes were basically a viewing tube with one lens, like a magnifying glass, placed inside. As with microscopes today, an observer would place their eye on the top and their sample underneath the tube. The first microscopes developed could probably only magnify a sample's image tenfold.

THE COMPOUND MICROSCOPE

Many European researchers took these new microscopes and looked at tissues and other biological samples, some making their own improvements to the microscope. Robert Hooke, experiments curator at the Royal Society of London, is generally acknowledged as being the first to observe cells. In about 1660, he improved the existing single-lens microscope design by placing a second lens in the viewing tube. When appropriately placed relative to the other components, the second lens allowed for much greater magnification of a sample; he created the first compound microscope. Modern compound microscopes can magnify a specimen 1000 to 2000 times.

THE DISCOVERY OF CELLS

Using his newly developed microscope, Hooke made many detailed drawings of all sorts of biological objects around him, from insect parts to fungus to feathers, and much more. In 1665, he published his drawings in his book *Micrographia*. His most celebrated drawing from *Micrographia* is a thin section of cork, made from oak trees. This drawing is shown below. In the thin slice Hooke observed, he found it to be "porous, much like a honey-comb." He named the tightly-packed cavities "cells," after the Latin word "cella," meaning "a small room."

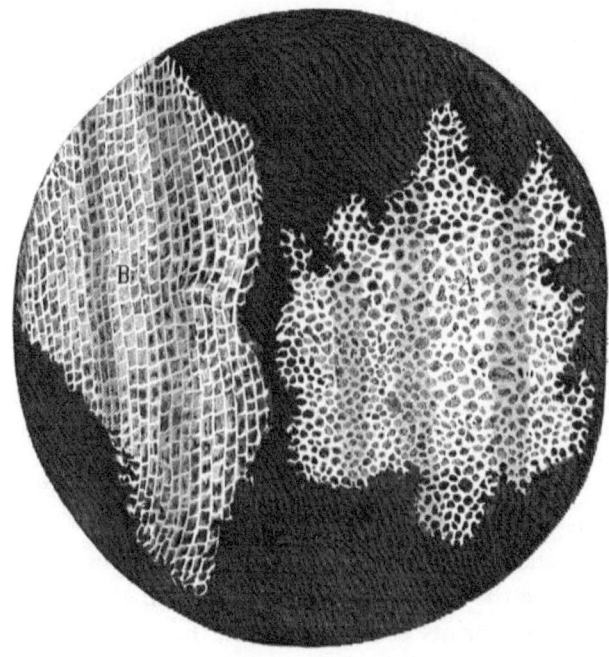

In 1665, Robert Hooke published this drawing of a thin section of cork that he observed under a microscope. It is where the biology-related word "cell" comes from.

Around the same time as Hooke, in the Netherlands, Anton van Leeuwenheok was also busy improving upon the original microscope designs, which led him to be the first to observe living cells. Van Leeuwenhoek was a cloth merchant, and to count threads he used magnifying lenses that magnified the cloth approximately threefold. He developed new ways to polish and grind the lenses; he increased the existing single-lens microscope's magnification to 270-fold magnification. With his improved microscope, in 1673 he was the first to see living organisms in pond water. While he called them "animalcules," they're now known to us as microorganisms.

Several other researchers during the 1600s were also looking at various biological specimens under microscopes, and this great demand soon made it quite apparent that even better microscopes were needed. Even with improvements, many imperfections in the microscopes could produce misleading observations; it was often difficult to know whether researchers saw actual cells, or just the result of aberrations, causing something to look like cells. Recognizing the need, lens markers throughout the 1700s worked to reduce aberrations; at the beginning of the 1800s, their efforts finally paid off with the creation of "achromatic lenses." However, these lenses still had problems and, throughout the century, they were improved upon to create "apochromatic lenses," which have even fewer aberrations.

THE NUCLEUS

As views of what was going on in these cells improved, scientists developed many theories for how the cells functioned and consequently got into many heated debates. One of the first was over the function of the cell nucleus. We know now that the nucleus is where the cell keeps its DNA, and that the DNA makes up the genetic blueprints for the cell (and the organism as a whole). Discovered in 1833 by Robert Brown, it's not too surprising that the nucleus was the first organelle (specialized cell component) found; it is a large, round cell part, often near the cell's center, and, importantly, it is relatively opaque, making it visible even with poor microscopes.

Mattias Schleiden, a German botanist and microscopy master, long examined plant cells under the microscope. Because plant cells have very rigid cell walls, they appear very different than animal cells, which lack walls and vary greatly in appearance from tissue to tissue. Consequently, Schleiden long thought that plants were made of cells, but animals were not. However, because Schleiden also came to believe that the nucleus was the most important and defining cell part, when his associate, the German zoologist Theodor Schwann, found similarities between plant and animal cells, such as the presence of a nucleus, he came to accept that all living things were made of cells. In 1837, this led to Schwann being the first to conclude that cells were "basic units of life."

While others generally accepted that cells were life's basic building blocks, the debate turned to how these cells were created. Schleiden believed that the nucleus generated a new cell around it in layers in a purely mechanical manner, much the way a crystal forms from around a small seed crystal (or, confusingly, "nucleus"). However, others were already describing cell division, the process by which the nucleus and surrounding cell duplicate their material then split into two cells; their observations went against Schleiden's nucleus theory.

Once again, the scientific debate was resolved by the availability of improved tools. In the decades after 1850, methods of preparing samples for microscope viewing were developed, aided by the increasing availability of dyes that colored only specific parts of the cell. Carmine red, the first stain, had been used for centuries as a dye in Mexico, where it was made by crushing chochineal insects. Carmine red selectively stains the nucleus, making it much clearer to see microscopically than before. Such advances undoubtedly contributed to Rudolf Virchow, a Germany physician, settling the cell origin debate in 1855: He reported, based on his observations, that all cells come from the division of other existing cells, not just from a cell nucleus.

THE CELL THEORY

Schleiden, Schwann, and Virchow developed the three fundamental parts of the cell theory which is still taught today: (1) All living things are made up of one or more cells. (2) Cells are the basic units of biological structure and function. (3) Cells can only come from preexisting cells.

Armed with this basic knowledge of cell biology, further investigations into the intimate, complex workings of cells were later made possible with the use of more advanced microscope techniques, specific dyes, and other tools that developed, such as ultracentrifugation. These tools allow scientists to make new discoveries to this day. For example, when I was a graduate student at the University of California, Santa Barbara, I had access to the many powerful microscopes, including electron microscopes that can magnify the image of a sample to around 500,000-fold (allowing resolution of individual proteins and DNA molecules), that allow scientists to conduct their research. Such tools allowed researchers to greatly understand the many different cellular components, or organelles, throughout the 20th century. These organelles, and their stories, will be discussed in Part II.

Further Reading

The following resources may be of interest for those wanting to learn more about the early history of the cell:

- William Bechtel's book *Discovering Cell Mechanisms: The Creation of Modern Cell Biology*
- Arthur Hughes's book *A History of Cytology*
- Henry Harris's book *The Birth of the Cell*

History of the Cell, Part II

Since the mid-1800s, collaborative efforts between have led to incredible progress in understanding how cells work

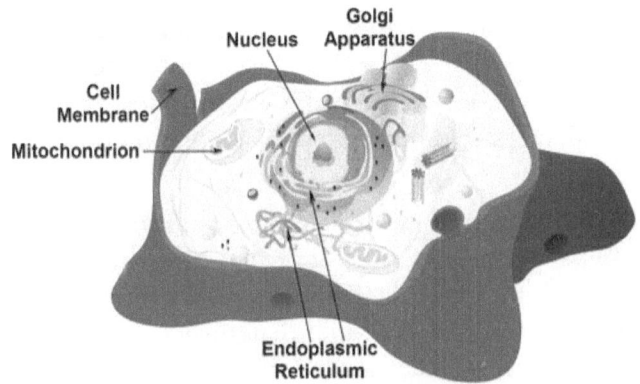

A simplified version of the cell, identifying some key organelles: the nucleus primarily directs what proteins need to be made, which are created in the endoplasmic reticulum and shipped to the Golgi apparatus for distribution throughout the cell, all the while mitochondria provided the cell its energy.

Our understanding of the biological cell has come a long way since Robert Hooke first saw it under the microscope in 1665, and most of this rapid progress has been in the past 150 years. Before then, cell scientists were just determining that cells are present in all living things and the role of the first known organelle, the nucleus, as explored in the previous section.

While continuing to better understand the nucleus, researchers also started to investigate other organelles in the cell. Inside each cell are many compartmentalized subunits, each serving a different function in the machinery of the cell; these subunits are called organelles. As vital as the microscope was to first seeing and understanding the cell, many other tools served key roles in elucidating the crucial functions of organelles.

THE CHROMOSOMES

Although the nucleus was known to split in half when cells divide, it was unclear what its exact role was. Only the advent of dyes that selectively stain the nucleus and better methods of preserving samples allowed German biologist Walther Flemming, in 1879, to see tiny "rods" within the nucleus. Due to their visibility, he named them "chromatin" (the prefix "chroma" means "color"). These structures are what we know today as chromosomes, which are tightly packed, organized bundles of DNA within every nucleus: the blueprints for life.

Observing the chromosomes, Flemming studied how the nucleus functions when the cell divides into two new cells. In 1882, he reported that during cell division the chromosomes align in the middle of the nucleus and split lengthwise, becoming equally distributed into the two new cells. However, it was not until the early 1900s that the role of the nucleus in hereditary was recognized, after genetics studies using pea plants, done by the Austrian priest Gregor Johann Mendel in the 1850s and 1860s, were rediscovered. Mendel is posthumously known as the "father of genetics."

We now have a much better understanding of not only how the chromosomes in the nucleus of every cell store a person's genetic information, and how these genetics are passed on to a new cell or the next generation, but also why, for example, liver cells know to be liver cells and don't change into brain cells. Although every cell in a body shares the same genome, factors in the nucleus control which genes a given cell uses, controlling the cell's identity and telling it how to function.

THE CELL MEMBRANE

At the same time, researchers started investigating another central feature of the cell: its surrounding membrane. In 1773, the English surgeon William Hewson reported seeing red blood cells shrink and swell depending upon the liquid they were in and proposed that cells were envelope-bound bodies. This is because membranes readily allow water to go inside the cell or leave it (using a process called osmosis). Unfortunately, Hewson's work was largely unknown and because the cell membrane was difficult to see with the microscopes of the time, there was much debate over whether it existed for some time.

The membrane dilemma was settled at the end of the 1800s by Charles Ernest Overton, at the University of Lund, Sweden. Overton repeated Hewson's findings, but went further and showed that the membrane is semi-permeable; cells actively select specific molecules to bring inside the cell, and others to transport outside. The membrane was quite real, and more excitingly, it was active! Substantial progress in understanding the membrane was made during the first half of the 1900s. We now understand it has a key role in not only regulating what enters and exits the cell, but also in maintaining the cell structure and communication between cells.

THE GOLGI APPARATUS

At the turn of the 20th century, all the pieces were in place for researchers to focus on what was going on in the rest of the cell. After the nucleus, one of the next organelles identified was the Golgi apparatus, seen by Italian scientist Camillo Golgi in 1898. As happens often in science, Golgi wasn't necessarily the first to observe the structure; several other researchers had reported similar structures, but because Golgi was very consistent and clean in his methods and results, his findings were the first widely-accepted. Consistent results were especially difficult because the Golgi apparatus actually varies in shape, making many think it was an artifact of poor microscopes and stains for a long time. But in the 1930s and 40s, researchers established its role in cell secretion. Today, we know the Golgi apparatus is not only involved in secreting large molecules, but also in protein modification. Overall, the Golgi apparatus determines where proteins need to go in the cell, AND packages and delivers them, like the cell's own post office.

THE ULTRACENTRIFUGE AND
THE MITOCHONDRIA

By 1940, the study of cells had come as far as it could with just the microscopes. Biologists increasingly turned to biochemistry to answer cellular questions. Additionally, around the 1930s a revolutionary investigative tool became available: the ultracentrifuge. The ultracentrifuge (a high-speed centrifuge) can separate particles by size and weight, as it spins samples at very high speeds, up to 1,000,000 times the force of gravity in modern machines.

In 1933, Robert Bensley, at the University of Chicago, used the ultracentrifuge to separate specific organelles from the rest of the cell to better study them. Bensley isolated the mitochondrion (plural mitochondria), which had been recently discovered by German pathologist Richard Altmann. Bridging the gap between seeing the cells under the microscope and understanding their biochemistry, Bensley confirmed his centrifuged product through staining. By the 1950s, researchers were widely applying ultracentrifuge techniques to determine the function of different organelles.

But despite knowing that mitochondria were present, their role in the cell remained murky. This understanding was provided by Albert Claude, at the Rockefeller Institute, who collaborated with biochemists to measure the enzyme activity of the isolated mitochondria. (Enzymes are proteins that help cause specific chemical reactions to take place in the cell.) Mitochondria were found to have important oxidative enzymes, implicating their role in creating energy for the cell. In 1948, Claude accurately described them as "the real power plants of the cell." Fascinatingly, mitochondria are also thought to have once been bacteria that became absorbed by the host cell, an idea called the endosymbiotic theory.

THE ELECTRON MICROSCOPE AND THE ENDOPLASMIC RETICULUM

Another ground-breaking instrument of the time was the electron microscope. First available in the 1940s, it focuses a beam of electrons the way a normal microscope focuses a ray of light. Today's electron microscopes can magnify a sample's image 500,000-fold, but they are limited in the thickness of tissue they can penetrate. In 1945, Claude used an electron microscope to look at mitochondria, but he could not see them clearly as they were too thick. As any professor would do, Claude quickly set the task before the students in his lab.

Keith Porter, a postdoctoral student who joined Claude's lab in 1939, practically lived under the electron microscope, improving techniques to view cells. In the 1940s, Porter examined a "lace-like reticulum" structure at great lengths, and enhanced ways to image it with the electron microscope. Others were also studying the structure at the time and thought it might be involved in creating proteins. In the 1950s, it was given the name we know it by today: the endoplasmic reticulum. The endoplasmic reticulum is indeed the site where many proteins are made in the cell. It receives its instructions from the nucleus, which it surrounds, and sends the completed proteins to the Golgi apparatus, adjacent to it, for further processing and distribution.

Although not even recognized in the 1940s, by 1970 cell biology was a well-established scientific discipline. The picture of the cell as a functioning machine was coming together: The nucleus directed what proteins needed to be made; the proteins were created in the endoplasmic reticulum and shipped to the Golgi apparatus for distribution throughout the cell; all the while mitochondria provided the cell its energy. Of course, this is a great simplification; there are many other organelles and processes involved. The cell is so complex that researchers today are discovering new details all the time. At the University of California, Santa Barbara, where I conducted my graduate research, several research groups study not only the function of cell organelles, but the many molecular communication pathways that are constantly active in every cell.

Further Reading

The following resource may be of interest for those wanting to learn more about the more "modern" history of the cell:

- William Bechtel's book *Discovering Cell Mechanisms: The Creation of Modern Cell Biology*

Model Organisms

Better understanding ourselves through others

Studying this little fruit fly (*Drosophila melanogaster*) for the last century has allowed researchers to better understand genetics and developmental principles applicable to ourselves.

Researchers have long studied other organisms in hopes of better understanding the human body. The severe ethical issues raised by the prospect of using humans beings for biological studies, such as for studying the function of disease and the process of development, are self-evident enough, and this is why "model organisms" have been vigorously pursued as an alternative by researchers. There are many qualifications such an organism must have: it must be similar enough to humans, in some way, that the knowledge gained is applicable to us; it usually needs to be "simpler" than us so that the process being studied is easier to figure out; it must be easy to grow in a laboratory setting, preferably in great numbers and relatively quickly. There are many organisms over the years that researchers have found to fit this description. These "models" have become the center of attention in many laboratories throughout the decades, and have brought us a much greater understanding of ourselves and how we are all related.

By the late 1800s, researchers realized the importance of using animal models, as many different factors began to fall in place. Until this time, many scientists had been trying to simply apply their knowledge of physical laws to understand physiology, but this was not working out very well. At the same time, with the advent of mass production, standardized equipment was becoming widely available to researchers. In 1900, attention was also brought back to the field of genetics, as the research done by Gregory Mendel, the "father of genetics," in the 1860s was rediscovered. Researchers were soon looking for the best animals in which to study this novel field of genetics.

THE FLY

While many would look upon the tiny fruit fly as a pest, scientists saw it as a great opportunity. The fruit flies (*Drosophila melanogaster*) have many attractive features: they are small, taking up little space in the laboratory and being relatively cheap to maintain; they reproduce quickly and in great numbers (a mature female can lay up to 100 eggs a day, and one fly goes through its lifecycle in about 10 days); and like humans, they are eukaryotes (which includes all animals, plants, fungi, and some single-celled organisms, with the defining characteristic being the presence of a nucleus), although fruit flies are much "simpler" than we are (with only 4 chromosome pairs to our 23 pairs).

During the first decade of the 1900s, a few research groups decided to try these new models out in their laboratories, and word of their appeal quickly spread; the fruit fly became the first widely accepted model organism in laboratories. Thomas Hunt Morgan at Columbia University was one of the first to get in on the fruit fly craze. He decided it would be a perfect model to study genetics, as there were a variety of obvious, physical mutations different fruit flies had and passed on to their offspring, including different eye colors, body colors, and wing shapes, to name a few. These mutations could all be easily observed with a magnifying glass. He created his famous "Fly Room" to map these different physical mutations to separate locations on the fruit flies' chromosomes. These efforts were spectacularly successful and consequently he helped make the fruit fly a prominent genetics model for researchers in the 1910s and 1920s, and was awarded a Nobel Prize in 1933 for showing that chromosomes carry genes, a concept central to our understanding of heredity.

However, as the tools available to researchers improved, so did their desire to investigate smaller and smaller components of the biological world around them. When the electron microscope became available to researchers in the 1930s, it drastically changed the focus of research; they could now magnify an image of a sample 500,000-fold. The previously obscured world of bacteria and viruses was no longer so hidden from view, but was still far from understood.

THE BACTERIA

By the 1950s and 1960s, many groups were using the stomach bacteria *Escherichia coli* (*E. coli*) and bacteria viruses ("bacteriophages") to understand genetics on a smaller, molecular basis. Rarer genetic mutations were studied using bacteria, as they can be quickly grown in huge numbers (*E. coli* doubles its population size about every 20 to 30 minutes under ideal conditions). *E. coli* can also be manipulated to take in foreign DNA and make more of it as if it were its own, or even turn the foreign DNA into protein; *E. coli* is now commonly used in laboratories to produce large quantities of DNA and protein that are from other organisms. For example, *E. coli* produces the vast majority of insulin used to treat diabetics in the world today. Although bacteria are prokaryotes (while humans are eukaryotes), researchers also use bacteria to study cellular processes that are highly conserved (very similar) between prokaryotes and eukaryotes, such as gene regulation and proteins that play roles in disease. As examples of this kind of research, at the University of California at Santa Barbara (UCSB), where I was a graduate student, Christopher Hayes's laboratory investigates the function of ribosomes (the cellular machinery that creates proteins) in *E. coli* and David Low's laboratory studies infectious *E. coli*. To find out more about their research, see

http://www.mcdb.ucsb.edu/people/faculty/hayes or

http://www.mcdb.ucsb.edu/people/faculty/low.

THE MOUSE

While researchers learned a great deal from simple organisms like the fruit fly and *E. coli*, model organisms that are closer to humans were needed for many investigations. This is how the mouse came to the laboratory spotlight. Mice (*Mus musculus*) are very similar to humans not only genetically, but also physiologically; they can develop cancer and similar diseases as humans do, or be manipulated to develop diseases and carry genetic mutations. Additionally, mice have a relatively short lifespan and can reproduce in large numbers. Perhaps the researcher who is most responsible for the abundance of mice in laboratories today is Clarence Cook Little, who developed the first inbred mouse line in 1909 and went on to found The Jackson Laboratory in 1929 as a mouse cancer research facility. The Jackson Laboratory created many lines of mice for researchers to use globally, "normal" mice as well as mice that functioned as models of human diseases or cancers, ensuring that the mice used in laboratories around the world would be standardized. With the genome of the mouse sequenced in 2002, today these well-characterized mammals are often used in laboratory work as research transitions to the clinic, frequently in cancer and tumor research.

THE WORM

One of the most widely used model organisms was a relatively late arrival to the scene. The appealing simplicity of the nematode (*Caenorhabditis elegans*, or simply *C. elegans*) prompted Sydney Brenner (Medical Research Council in Cambridge, UK) to use it and spread it to research laboratories in the 1960s. "The worm," as it is fondly referred to by researchers, is an ideal model organism; among other appealing traits, it's small (only 1 millimeter-long), can reproduce quickly and in great numbers (by self-fertilizing with hermaphrodites or introducing new genetics with rarer males), and is transparent, making it easy to see through under a microscope. Brenner was attracted to *C. elegans* because he wanted to study how the nervous system develops and thought that the essential processes would be similar between humans and the more "simple" multi-cellular worm. Brenner created a diagram of all of the neurological connections within the worm, and for his work received a Nobel Prize in 2002. The appeal of using *C. elegans* for studying development was soon shared by many other researchers, some who went on to map the destiny of every cell in the egg; all of the approximately 1000 cells that make up the adult worm can be traced to precursors in the egg. As a high school student and later as an undergraduate student at the University of Colorado, Boulder, I was fortunate to do some research in William B. Wood's laboratory, which studies how the different cells in *C. elegans* have their fates determined. To find out more about his research, see http://mcdb.colorado.edu/directory/wood_b.html. At UCSB, Joel Rothman's laboratory also works on elucidating the intricacies of *C. elegans'* development to better understand the process of development in general. For more on his research, see http://www.mcdb.ucsb.edu/people/faculty/rothman.

THE RETURN OF THE FLY

Though the fruit fly (*D. melanogaster*) was heavily used in genetics research during the first couple decades of the 1900s, it was too complex a model to use for more detailed understanding of genetics at the time, and so it faded from the research spotlight for a while. In the 1970s it made a big comeback, as the field of developmental biology began to take off. The many, well-characterized mutant lines of fruit fly that had been used for genetic studies were now used in developmental studies; researchers could trace a specific, obvious mutation back to its developmental origin, and thereby learn about the cell lineages during development. As examples of research being done today using the fruit fly, at UCSB Stephen Poole's laboratory uses the fruit fly to study genetics during development, and William Rice's laboratory uses it in evolutionary genetics studies. To find out more about their research, see http://www.mcdb.ucsb.edu/people/faculty/poole and http://www.lifesci.ucsb.edu/eemb/faculty/rice/research/research.html.

THE OTHERS

Several other model organisms are used in laboratories around the world. At UCSB, for example, this includes a variety of plants (to study plant growth, fruit ripening, and host-microbe symbiosis, among more), and animals that play key roles in ecological systems. The proximity of UCSB to the ocean has allowed it to develop extensive marine laboratories, perfect for the study of many types of unique marine organisms. This has allowed researchers to better understand some unusual stem cells (such as in the marine primitive chordate organism *Botryllus schlosseri*), investigate natural biomaterials (such as what mussels use to stick to underwater surfaces), and study the process of fertilization and early development in the sea urchin (*Strongylocentrotus purpuratus*), which is a very similar process to humans.

In addition, there are other model organisms that UCSB and other research universities use that do not need such specialized conditions as these marine critters. For example, before starting my graduate studies, I had the pleasure of working with a relatively simple eukaryote, the single-celled, freshwater ciliate *Tetrahymena thermophila*, which is shown below. This was in the laboratory of Eduardo Orias at UCSB, where it is used as a model for experimental cell and molecular biology. To find out more about Orias, who has worked at the university for more than 50 years, and his research see http://www.mcdb.ucsb.edu/people/faculty/orias.

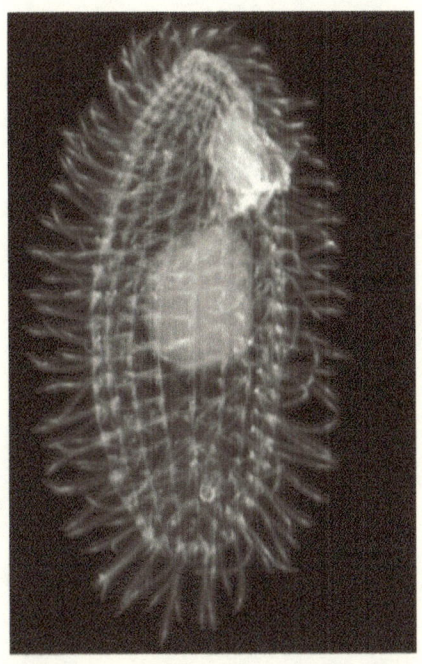

Tetrahymena thermophila is a relatively simple eukaryote (it is single-celled) that is often used in laboratories.

While not an "organism," perhaps the most frequently used biology model today is cultured cells. (These are cells that were originally from an animal, but are grown separately, outside of the animal.) This allows researchers to study a wide variety of processes in human cells. Additionally, it bypasses many of the ethical concerns associated with using animals in studies. The field of stem cells, which was discussed in detail in Chapter 1, is perhaps at the forefront of using cultured human cells to better understand human processes such as development, cancer, and wound repair.

Using different models to answer different questions, scientists at UCSB and institutions around the world spend their days examining the delicate inner workings of life. The models may look different, but they all contribute a piece to the puzzle of understanding ourselves and our world.

Further Reading

The following resources may be of interest for those wanting to learn more about model organisms:

- Angela N. H. Creager, Elizabeth Lunbeck, and M. Norton Wise's book *Science Without Laws: Model Systems, Cases, Exemplary Narratives*
- The Jackson Laboratory's website at http://www.jax.org
- The National Institutes of Health's webpage "Model Organisms for Biomedical Research" at http://www.nih.gov/science/models

In Vitro Fertilization, Part I: A Brief History

Accounts for over one percent of U.S. births

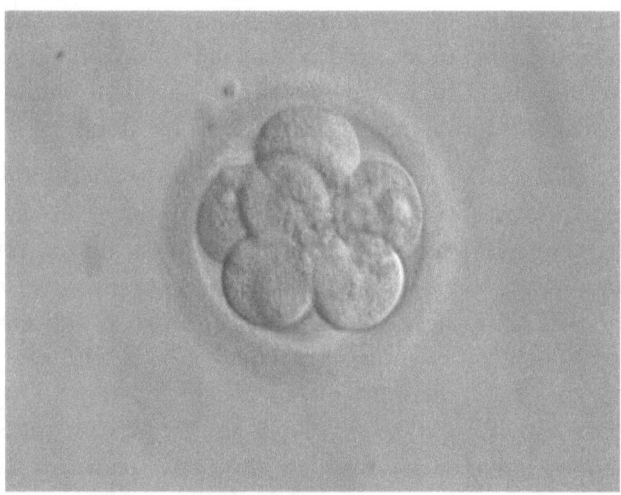

During *in vitro* fertilization, embryos that have only eight cells, such as the one shown here, are implanted into the uterus of a woman.

Infertility is an extremely prominent issue for many couples. One solution that increasing numbers of people turn to each year is *in vitro* fertilization (IVF).

INFERTILITY

Infertility is a very real, very silent, and very chilling health issue that many couples are forced to confront. One in ten people in the United States will face fertility problems. For a woman, these issues are closely linked to her age: When a woman is in her late 20s, her fertility starts to decline. It falls even more quickly when she turns 30, and continues to decrease by 3 to 5 percent each following year. It falls even more quickly after she's 35, and plunges when she becomes 40. Combined with male fertility problems, by the age of 30, seven percent of couples are infertile, and this increases to 33 percent by the age of 40.

These statistics combined with the strong desire to have one's own biological child drive about one million people in the U.S. to seek fertility treatment of some kind every year, and some 300,000 resort to IVF. For many, it works. Each year, over 1 percent of the babies born in the U.S. are through IVF. That's around 50,000 births. The industry of assisted reproductive technology (ART) is booming. Fertility drugs bring in about $3 billion annually, and there are over 500 fertility clinics in this country alone.

But how well does IVF actually work? The most successful fertility clinics report over 30 percent pregnancy rates. However, this means that some 60 percent of the trials, at least, do not result in a pregnancy. That success ratio is disappointingly low to many couples, especially in light of the costs both fiscal and emotional. IVF is very expensive, sometimes costing over $10,000 for one treatment, making it simply out of reach for most lower- and middle-class couples. And, because of the 30 percent success rate, many couples have to go through repeated treatment cycles to become pregnant, and many give up. It can often be a treadmill: the hopeful couple keeps trying, thinking that the next cycle will be the one that brings a child into their lives.

But for many it does work, giving them a biological child of their own which they could probably otherwise not have, using an approach that was probably not available to their parents.

HOW DID IVF COME TO BE?

Although IVF has only recently become widely used to treat infertility problems, people have been seeking medical help to overcome reproductive barriers for a long time.

In 1884, the first known artificial insemination using donated sperm took place – and the woman inseminated did not even know what had happened. A couple having fertility problems were examined by Dr. William Pancoast of the Jefferson Medical College in Philadelphia. After finding that it was the husband's sperm that was the problem, and the woman's reproductive system was fine, the doctor realized their fertility problems could be solved using donated sperm through artificial insemination. This had not been done in humans previously, and such an idea at the time would have been rather scandalous, possibly even considered to be an act of adultery.

Consequently, the woman, and most likely the husband as well, were not told what was causing the fertility problems, or what the "treatment" would be. She was only told to return for a follow-up appointment to have "her" fertility problems fixed. When she arrived, she was knocked out with chloroform and the doctor chose a sperm donor from his medical students that were present. The student masturbated into a cup and the semen was injected into the unconscious woman's uterus. It was successful: Nine months later she gave birth to a baby boy, still not knowing how she became pregnant.

Not until the 1930s did the idea of artificial insemination receive much publicity. During the 1940s, researchers found how to successfully freeze donated sperm, which later led to many sperm banks opening in the 1970s. However, artificial insemination is only beneficial when the infertility of a couple is due to the "male factor." If the woman is having reproductive problems, the situation can be much more difficult to solve, although this was greatly helped with the development of pregnancy-stimulating hormones in the 1960s. However, because such drugs can only help so much, the arrival of IVF revolutionized the existing field of assisted reproductive technology.

Like all science, IVF didn't come out of nowhere; researchers had been performing the studies that enabled the development of IVF for around half a century. In England in 1890, Walter Heape transferred embryos from one rabbit to the reproductive tract of another rabbit, and successfully had the recipient rabbit produce offspring using this method. For decades, more animal studies built upon Heape's original findings. But scientific research cannot progress without the proper tools. Techniques for culturing cells improved during the 1960s and 1970s and, together with a better understanding of the hormones released during pregnancy, this allowed researchers to eventually culture embryos outside of the human body.

In 1978, the first person was born from IVF. Lesley and John Brown had been trying for nine years to have a child, but one of Lesley's fallopian tubes was blocked and consequently her eggs could not travel down to the uterus to be fertilized. On their mission to have their own biological child, Mr. and Mrs. Brown met Drs. Robert Edwards and Patrick Steptoe, who had been working on IVF treatments. Previous couples had tried IVF, but none of the treatments resulted in successful pregnancies. However, the IVF treatment worked for the Browns and on July 25th, 1978, Louise Brown was born. Many feared that a child born through IVF would be abnormal, or ostracized by society, but as Louise was shown to be a healthy, happy baby, and accepted by others, these fears mostly subsided. As of writing this, Louise Brown is still doing quite well; she has a naturally conceived child of her own and will turn 36 in 2014.

SO WHAT EXACTLY IS IVF?

In essence, in IVF sperm and eggs are removed from donors and fertilized outside of the body (in a Petri dish) and then implanted back into a woman's uterus. (The term "in vitro" literally means "within the glass" in Latin and is often applied to the study of a normal biological process replicated outside of an animal. In the case of IVF it is a bit misleading as the embryos are not placed in a glass tube, but in a plastic Petri dish.)

One treatment, or "cycle" of IVF, takes about four to six weeks from start to finish. Before the eggs can even be removed from the woman, her ovulation cycle is strictly regulated using drugs that first prevent ovulation (for two weeks) and then drugs that stimulate ovulation (for about ten days). This helps ensure that she will have a large number of eggs ready to be harvested at a given time. As her ovulation cycle is being regulated, her developing eggs are monitored through ultrasound and her hormone levels are observed through blood samples. About 34 to 40 hours before the eggs are to be harvested, the woman is given a final drug to "ripen" the eggs. The woman is then anesthetized and a fine needle is guided to her ovaries and her eggs are removed. The eggs are mixed with the sperm from the husband, partner, or donor. If all goes well and the eggs become fertilized, the resultant embryos (fertilized eggs) start growing in a Petri dish in the clinic.

Only a few days after the eggs were removed, the best embryos are selected to be implanted; the embryos have about eight cells at this point. A thin catheter is used to insert the embryos, usually one or two, back into the uterus. (This is easier than removing the eggs; no anesthetic is used.) Two weeks after implantation, the woman undergoes a pregnancy test; if it is positive, the pregnancy should progress normally.

INTRACYTOPLASMIC SPERM INJECTION

While IVF has helped many couples conceive their own biological children, the "male factor" was still often a problem. Many couples bypassed this by using donor sperm, but this meant that the "recipient father" would not be the child's biological father. To help satisfy men's desires to more easily be biological parents, a procedure called intracytoplasmic sperm injection (ICSI) was created in the early 1990s. Previously, in order to fertilize an egg in IVF, hundreds of thousands of sperm were required. However, for ICSI to work it theoretically needs only one functional sperm. In ICSI, a single sperm is taken and inserted directly into the egg, meaning that extremely low sperm counts are no longer necessarily a significant barrier to having children.

The ICSI technique has become so popular that in 2005 ICSI was used in 60 percent of the IVF treatments, even though only 37 percent of the IVF cases had a problem with the "male factor." ICSI has dramatically changed the fertility industry. The success of ICSI has had other consequences. It's believed that heterosexual use of sperm banks is declining. However, the use of sperm banks by single women and lesbian couples is actually increasing.

CONSEQUENCES OF IVF

Now that people have access to the amazing reproductive tool of IVF (and often in conjunction with ICSI), what are the consequences, socially, scientifically, and ethically? These topics will be explored in the next section, which will look at the legalities surrounding the practice of IVF, the isolation of embryonic stem cells, the creation of "designer babies," the occurrence of multiple births (twins, triplets, and more!), concerns of a growing infertile population, and the growing practice of surrogacy.

Further Reading

The following resources may be of interest for those wanting to learn more about IVF:

- Ruth Deech and Anna Smajdor's book *From IVF to Immortality*
- Naomi R. Cahn's book *Test Tube Families: Why the Fertility Market Needs Legal Regulation*
- Elisabeth Hildt and Dietmar Mieth's book *In Vitro Fertilisation in the 1990s*
- Centers for Disease Control and Prevention's webpage "Assisted Reproductive Technology (ART): What is Assisted Reproductive Technology?" at http://www.cdc.gov/art/

In Vitro Fertilization, Part II: Consequences

One in a hundred U.S. babies

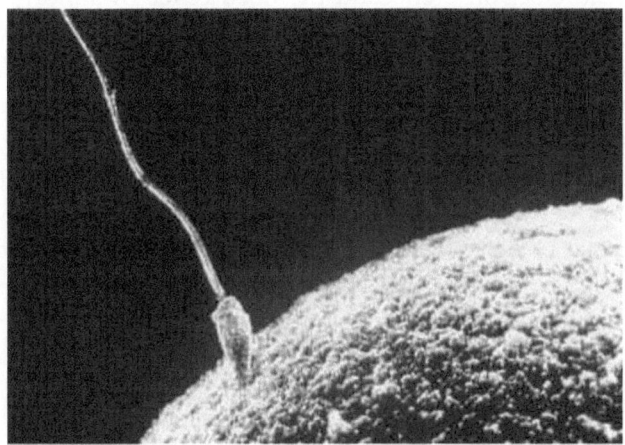

During *in vitro* fertilization, sperm and eggs are united (as shown above) in Petri dishes to create embryos that are implanted into the uterus of a woman.

The technology has come a long way since *in vitro* fertilization (IVF) was first successfully used in humans, resulting in the birth of Louise Brown in 1978.

Today, over one percent of the births in the United States are due to IVF. However, the practice has also gained a lot of baggage along the way. Here, we will explore the many social, legal, and economical consequences of widespread use of IVF.

In the previous section, we explored the history and biology of IVF, which will be summarized here. IVF is often used by people who desire to have their own biological children, but are unable to do so due to fertility problems. In IVF, sperm and eggs are removed from donors and fertilized outside of the body, in a controlled laboratory environment, and then the fertilized eggs are implanted into a recipient woman. This is, of course, an oversimplification: A variety of options can be available to someone interested in IVF, from what to do with extra embryos, to selecting for certain genetic traits, to choosing to have the pregnancy "outsourced," and much more.

FROZEN EMBRYOS

Embryos have been frozen and successfully used in pregnancies since the early 1980s. While there is a slightly lower pregnancy rate, there are no clear differences in the occurrence of abnormalities in children born from frozen versus fresh embryos. Why would a couple choose to freeze their embryos? If the first IVF cycle doesn't work, freezing makes it quicker to try a second cycle; or the embryos could be used much later to potentially create another child. However, because so many couples freeze their embryos, there are now over half a million frozen embryos in the U.S. Many wonder what to do with these embryos, which could be donated to another couple or used for valuable research, such as the creation of embryonic stem cells.

EMBRYONIC STEM CELLS

In 1998, researchers led by Dr. James Thomson (who holds a faculty appointment at the University of Wisconsin and at the University of California, Santa Barbara, where I did my graduate research) first isolated human embryonic stem cells (hESCs), which were discussed in Chapter 1. This was done using early-stage embryos (four or five days post-fertilization). Because hESCs can become virtually any cell type, they have great potential for studying and curing diseases and injuries, and, theoretically creating organs.

These early-stage embryos were from IVF clinics and donated with full consent. The embryo contains approximately 150 cells at this point and is a hollow, fluid-filled sphere. Often embryos used to create hESCs would have otherwise been discarded by IVF clinics because they were abnormal, though it's since been shown that embryonic stem cells can be made from a small piece of a normal embryo, and that embryo can still normally mature.

DESIGNER BABIES

One of the most controversial recent developments of IVF is using it to select for certain genetic traits in embryos, and, consequently, children. How is this done? After sperm and eggs are removed from the donors, the sperm is allowed to fertilize the eggs and the resultant embryos (when they reach three days old) can have one cell safely removed to perform genetic testing on their DNA.

One form of genetic testing is called "preimplantation genetic screening" (PGS), which counts the number of chromosomes in the embryo's cell. If a woman is over 35 years old, her eggs have a significant chance of having genetic abnormalities. Specifically, a person normally has 23 pairs of chromosomes (for a total of 46 chromosomes), but embryos made from an egg with genetic abnormalities may have an extra copy of one chromosome, resulting in a "trisomy" (three copies of one chromosome instead of two copies). Or, instead of a pair of chromosomes they may have only one. Most of these anomalies are lethal, which can cause repeated miscarriages. An exception is Down Syndrome, which is caused by three copies of the chromosome labeled "21," and is why it is also called Trisomy 21. PGS allows couples to prevent implanting an embryo with any of these terrible genetic maladies.

Taking this a step further, IVF can allow prospective parents to select for, or against, specific genetic traits through "preimplantation genetic diagnosis" (PGD). One of the least controversial uses of PGD is screening against an embryo inheriting a high-risk genetic disorder from the parents. One of the many couples who have used PGD in this manner are Susan and Chris Paget Dunthorne in the UK. Mr. and Mrs. Dunthorne's first child died when he was only a few months old, having inherited cystic fibrosis from the couple; they had not known they were carriers. The couple wanted to make sure their next child did not have the devastating disease, and did so by undergoing IVF and having their embryos screened for this genetic condition. Eventually, Susan had a healthy son through IVF and PGD.

PGD has also been used to select the gender of an embryo, though mainly for preventing a sex-linked genetic disease from being transmitted. Social, rather than medical, reasons are usually what stops this practice. One clinic, The Fertility Institutes in the U.S., started advertising to do genetic screening for couples to select specific hair and eye color, as well as complexion and gender, but halted the campaign (except for gender selection) after much public backlash.

More controversially, PGD has been approved to create "savior siblings." Raj and Shahana Hashmi, from the UK, had a son, Zain, who suffered from beta thalassaemia, a blood disease with severe symptoms; he would require painful, life-long medical care or have a life expectancy of a few years. Because umbilical cord blood from a baby of the same tissue type could save Zain, Zain's parents decided to have another child using IVF, and find a match by screening for a human leukocyte antigen, which determines tissue type. While their choice was approved by governing bodies, the IVF attempts were not successful. The Hashmis continue to look for matching bone marrow donors for Zain.

LEGALITIES OF IVF

In the U.S., there is a striking lack of uniformity in laws and regulations surrounding IVF. The laws vary from state to state and there are very few federal laws. For donating eggs or sperm, the American Society for Reproductive Medicine has created guidelines to ensure safety tests are performed on the donated material, but that's it. Additional requirements for donors differ between states. There are great differences even between banks in the donor information provided to prospective recipients, as well as the requirements for recipients. For example, some require that recipients be married or heterosexual, while others do not.

Paternity questions are often raised. While people can create documents that waive the donors' parental rights, only a few states assure that the contracts are legally valid. Such parental claims are often resolved in private agreements or courts.

Laws have been gray over who can claim ownership of sperm or eggs to be used in IVF. As a result of multiple instances where sperm was collected from dead or comatose men whose spouses (or significant others) then wished to undergo IVF using the genetic material, the UK has made it illegal to remove, store, or use anyone's gametes in the UK without the donor's written consent.

Individuals seeking to use even their own gametes have come up against legal difficulties. For example, in 1999 24-year-old Carolyn Neill in the UK had cancer and found she would have to undergo chemotherapy. Because this would most likely leave her infertile, and she had not had children, Neill decided to have some of her eggs frozen. After the chemotherapy treatment was successful, Neill found that she could not retrieve her eggs because using thawed eggs in IVF had not been successful in the UK yet, and consequently such practice was illegal. However, fearing human rights litigation, the ban on using frozen eggs was eventually lifted. While success rate is still quite low, children born from frozen eggs have not been shown to have any more abnormalities than children born naturally.

MULTIPLE BIRTHS

One of the highest risks associated with IVF is having multiple children from one pregnancy. About 24 percent of all IVF births are twins or triplets. However, this can be regulated by controlling how many embryos are implanted in the recipient's uterus.

In the UK, usually one or two embryos are implanted, and a healthy woman of childbearing age has about a 28 percent chance of pregnancy from one IVF cycle. Because the successful pregnancy rate for older women is lower, women over 40 can have three embryos implanted at once, the legal maximum in the UK. (It should be noted that although a menopausal woman cannot naturally become pregnant, her uterus can still be prepared to successfully receive an embryo through IVF, leading to many publicized mothers in their 60s and 70s.)

In the U.S., the American Society for Reproductive Medicine gives similar guidelines, recommending that women under 35 should have one or two embryos implanted at once, and that women 35 to 40 implanted with three at most, bumped up to a maximum of five for women over 40.

However, even though the ASRM has "guidelines," the U.S. does not have any federal restrictions on the number of embryos that can be implanted in a woman during one IVF treatment. This led to 33-year-old Nadya Suleman, the famous "Octomom," having octuplets in 2009 (in addition to the six older children she previously conceived through IVF).

Multiple births are also associated with a higher risk of miscarriages. In 1996, Mandy Allwood, who had been taking fertility drugs, became pregnant with octuplets (though not through IVF). Despite the strong recommendations of doctors to undergo selective abortions to help some of the fetuses survive, Allwood continued with the pregnancy and all of the octuplets were lost in a miscarriage when Allwood was 19 weeks pregnant. Federal regulation of the number of embryos that can be implanted during IVF may help decrease the number of multiple births associated with it, which can result not only in miscarriages but in women having far more children than they'd planned for.

SURROGACY AND "OUTSOURCING" PREGNANCY

Similar to the legal problems surrounding IVF in general, laws surrounding surrogacy in the U.S. differ between states – especially the rights of surrogates, donors, and "intending parents." In some states surrogacy is banned. But where it is allowed, usually a contract is created between the involved parties, and when conflicts have arisen, the "intending parents" have usually ended up with the baby.

However, this is not always the case. One such instance involved a British woman who carried a baby for a Californian couple, using embryos fertilized through IVF. When the woman became pregnant with twins, and refused to have one fetus aborted, as the couple requested, the couple reneged on the contract and the British woman ended up with two new children.

Like many jobs, surrogacy is also increasingly becoming "outsourced." A booming surrogacy industry has taken root in India, where surrogates incubate babies for couples in European countries, the United States, or other countries. In India, a surrogacy that could cost up to $70,000 in the United States can cost only $12,000. At the forefront of this industry is the Akanksha Infertility Clinic in Anand, where pre-screened women are paid $5,000 to $7,000 for each pregnancy. Interested recipients can either have their own sperm and eggs used, or use donor gametes. For the well-being of all parties involved, many feel this outsourced practice needs to be much more tightly regulated, especially since different countries have very different laws on surrogacy.

POTENTIAL CONSEQUENCES FOR COMING GENERATIONS

While IVF allows people to have children, when they may not have been able to otherwise, adding a net gain to the global population, there are concerns that increased use of IVF may actually be propagating infertility genes. Infertility is often caused by genetic abnormalities, frequently in the male factor, which are easily solved by intracytoplasmic sperm injection, which was discussed in the previous section. However, it's generally recommended that discriminating against potential IVF clients on the basis of genetics be avoided, because of its eugenic undertones.

Many of the questions and concerns surrounding the multifaceted nature of IVF may be resolved in courtrooms as its use becomes more widespread. Perhaps eventual federal solutions will put interested parties more at ease for what to expect, helping smooth some of the bumps on the emotional rollercoaster.

Further Reading

The following resources may be of interest for those wanting to learn more about IVF:

- Ruth Deech and Anna Smajdor's book *From IVF to Immortality*
- Naomi R. Cahn's book *Test Tube Families: Why the Fertility Market Needs Legal Regulation*
- Elisabeth Hildt and Dietmar Mieth's book *In Vitro Fertilisation in the 1990s*
- Centers for Disease Control and Prevention's webpage "Assisted Reproductive Technology (ART): What is Assisted Reproductive Technology?" at http://www.cdc.gov/art/
- Medical Tourism Corporation's webpage "Low Cost Surrogacy in India" at http://www.medicaltourismco.com/assisted-reproduction-fertility/low-cost-surrogacy-india.php
- Akanksha Infertility Clinic's website at http://ivf-surrogate.com/ and Surrogacy Abroad Inc.'s website at http://www.surrogacyabroad.com
- ABC World News's article "More Americans Now Traveling to India for Surrogate Pregnancy" by Clarissa Ward at http://abcnews.go.com/WN/adoption-india-americans-plan-surrogacy-abroad/story?id=10487880

Genetically Modified World, Part I: Plants

Performance-enhanced plants

Corn is one of many crops that has been genetically modified to be pest-resistant or herbicide-resistant. It has become so successful that in the U.S. about 60% of corn grown is GM corn.

Advancing technologies are constantly changing every aspect of modern life, from how we treat our colds to helping us feel more energetic. Over the past two decades, this has extended to giving us the option of growing and purchasing foods that have been genetically modified (GM). As with all major advancements, GM crops have been met with resistance, although they offer a more efficient means of generating nutrient-rich produce and saving thousands of lives.

While many opponents of GM crops argue that they are "unnatural," this certainly isn't a new or unnatural human approach; we've been molding the plants around us to make them more useful

129

for our nutritional needs since prehistoric times, or for around 10,000 years. During this time, people have been selecting for certain desirable traits in plants, such as bigger fruits or pest tolerance, controlling what the generations of crops to come would look like. The largest difference in the current GM approach is the fast speed at which the selection process is occurring, which gives people relatively little time to understand potential wide-scale and long-term impacts.

In the early 1900s, the agriculture business completely changed with the advent of mass-produced farming machinery, pesticides, herbicides (to kill weeds), fertilizers, and the creation of more vigorous, higher-yielding "hybrid" plants. While this so-called "Green Revolution" created crops with unprecedented yields, it was met with trepidation by the public, just as the GM agricultural revolution that began in the 1990s is being met with today.

The number of GM crops around the world has skyrocketed since the first creations in the early 1990s; in 1996, they took up about 6,500 square miles, but by 2009 they covered nearly 520,000 square miles, spread across 25 countries, 16 of which are developing. While the industry had a very successful start in the U.S., it then slowed down globally as it encountered several hurdles, from showing that the industry is safe for people and the environment, to settling intellectual property disputes raised by the companies that made the GM crops, to gaining public acceptance. Fear of "Franken-foods" has driven many policies in the world concerning GM crops, and is why they must be labeled on shelves in countries in the European Union and elsewhere. But are these fears justified?

Genetically modified plants (also generically called genetically modified organisms [GMOs]) are plants that have had genes from other plants or organisms inserted into them. These "foreign" genes can make their new plant hosts have a range of desirable traits that make farming much more efficient, such as requiring less water, using less fertilizer, and/or making do with poorer soil. Two of the most widely used traits in GM crops are toxicity to insects, and resistance to herbicides. So it's no surprise that GM crops have been repeatedly shown to outperform "normal" crops. But the researchers haven't stopped there; other GM crops have been designed to produce high amounts of vitamins and potentially virus vaccines!

SETTING THE STAGE: THE FLAVR SAVR TOMATO

The first GM crop found its way to grocery store shelves in 1994. It was a tomato genetically modified to stay hard even when ripe. Why would this be useful? For a long time, commercially sold tomatoes have been harvested when still firm and green, and then treated with ethylene gas to ripen them. This is so they're not transported soft and damaged. So much for being "natural."

How did the GM tomato "FLAVR SAVR," as it was called, keep its youthful firmness while still becoming red and ripe? It comes down to an enzyme called polygalacturonase ("PG" for short). Made by ripening tomatoes, PG breaks down the structural supports in plant cell walls (specifically pectin, a substance whose structural stability is what makes jellies and jams firm). Researchers thought that if they could just stop the unripe, green tomatoes from producing PG, they could prevent them from becoming soft. And they were right. To achieve this feat, scientists used antisense RNA technology; they knew the RNA sequence that the tomato used to create PG, and so they introduced an interfering RNA sequence (an antisense one) that bound to the native sequence, preventing it from functioning.

But how did this antisense RNA find its way into the tomato plant? With a little help from nature. The researchers used a species of bacteria that normally attacks a variety of plants (specifically the bacteria *Agrobacterium tumefaciens*) but replaced part of the bacteria's infectious genome with the desired antisense sequence. When the tomatoes were exposed to this modified bacteria, the sequence found its way to the plant's genome. While "FLAVR SAVR" only lasted a few years on the market due to it not being profitable, this same strategy (and same bacteria) has been used again and again to create GM crops.

SOYBEANS: HERBICIDE RESISTERS

The second GM crops to arrive in supermarkets were GM soybeans. They were much more successful than the tomatoes, and today, GM soybeans are the most commonly used GM crop in the world. In 2005 in the U.S., nearly 90 percent of the soybeans grown were GM soybeans.

And how exactly have these crops been modified? Many have been made resistant to an herbicide called glyphosate (for you chemists out there, it's [N-(phosphonomethyl)glycine]), which is more commonly known as Monsanto's weed-killer "Roundup." Most commercially-sold herbicides are nonselective plant-killers, functioning by disrupting the plant's ability to carry out photosynthesis (vital to its survival). Some herbicides, like glyphosate, go so far as to destroy the plant's ability to make amino acids (which is necessary to make proteins in all animals). So you can imagine how making a crop resistant to a toxic herbicide like glyphosate can make things more convenient; spray the entire field with herbicides and the only thing left standing are the soybeans, or "Roundup Ready Soybeans."

But how do you make a plant resistant to glyphosate? Bacteria again come to the rescue. A mutant bacteria strain (*Salmonella typhimurium*) was found to resist the killing power of glyphosate, and researchers were able to isolate the gene responsible for this resistance and put it into the genome of soybeans, as well as several other crops. (Specifically, the gene encodes for a mutant version of the protein normally targeted by glyphosate; this mutant functions, but isn't recognized by the glyphosate.)

So what are the problems with spraying a field with general plant-killing herbicides? Possible herbicide-resistant weeds, for one. Just as bacteria can become resistant to antibiotics over time, plants can also have mutations which may cause a few herbicide-resistant weeds to become a "super-weed" epidemic. There is also some evidence that the glyphosate-resistant genes may jump from GM crops to weeds and other plants, but weeds don't need GM crops to become herbicide-resistant. While there's some question over whether using GM soybeans is actually more profitable, it's thought that the general simplicity of using Roundup Ready Soybeans has made many farming converts.

BT CORN

Next to soybeans, corn is the most commonly used GM crop worldwide. In 2005 in the U.S., about 60% of corn grown was GM. While some GM corn has been made to be herbicide-resistant, like GM soybeans, many have also been made to be pest-resistant or more nutritional. The manner in which GM corn has become pest-resistant is by borrowing a gene from, you guessed it, bacteria, specifically *Bacillus thuringiensis*, commonly known as "Bt." In the wild, this bacteria creates a small toxic protein that is released from the bacteria when it is eaten by an insect.

In the digestive track of the insect, the digested protein binds to the cells lining the gut and ruptures them, killing the insect. The toxin works well against many different kinds of insects, including the orders Lepidoptera (which includes the European corn borer, cotton bollworm, the tomato fruitworm, and other moths and butterflies), Diptera (flies and mosquitos), and Coleoptera (beetles, including corn rootworms).

Researchers isolated the gene from Bt that creates this toxic protein and put it into the genomes of many GM crops, including corn, tobacco, cotton, and tomatoes. While this successfully kills many very harmful crop pests, it has been criticized for also killing non-targeted insects and beneficial insects. However, proponents of Bt GM crops argue that using these crops is more environmentally sound than spraying massive amounts of pesticides.

COTTON: A SUCCESS IN DEVELOPING COUNTRIES

Cotton is the next most widely used GM crop globally after Bt corn. After being approved early on in the U.S., in 1995, it quickly made its way around the world. As of 2009, it covered nearly 62,000 square miles of fields (comprising 45 percent of all cotton acreage). It is mostly grown in India, China, and the U.S., and has been touted by some as a standard-bearer for the success of GM crops in reducing poverty in developing countries. Like Bt corn, the Bt cotton also has the Bt insecticide gene, allowing farmers to spend less money on large amounts of expensive pesticides. Bt cotton also has significantly higher yields (about 30 to 40 percent) than normal cotton, resulting in higher incomes than with normal cotton, which has helped to reduce poverty levels in developing countries that grow it, with particular success in India.

"GOLDEN RICE"

While the much-publicized "Golden Rice" is not available on grocery shelves yet, it is quite justified in receiving the attention it has. About 140 million people suffer from vitamin A deficiency, a serious malnutrition problem in poverty conditions, where diets consist of low-nutritional staple foods. Every year, nearly 3 million children die from vitamin A deficiency. This is why researchers created Golden Rice. Golden Rice is a GM rice that contains beta-carotene, the same chemical that gives carrots their orange color and which is a precursor our bodies can turn into vitamin A. (And it gives Golden Rice its distinctive yellow hue.) To create Golden Rice, researchers basically borrowed multiple enzymes necessary to make beta-carotene, from daffodils and a soil bacterium. Although Golden Rice is not sold commercially (as of 2012), it is estimated that it could save the lives of 40,000 children in India alone, every year.

WHAT DOES THE GM FUTURE HOLD?

The future of GM crops is exciting, if tumultuous. As just one example, already some GM bananas are being developed that may vaccinate against cholera or hepatitis B. However, GM crops have many hurdles in front of them which they must overcome to be widely grown and accepted. These range from biotechnology companies with intellectual property rights, to concerns about herbicide-resistant plants and possible effects on the environment, and consumers' health. To date, there are many conflicting results on whether GM crops cause health problems; it is still unresolved. The reality is that GM crops are here and it is our responsibility to make sure they are carefully evaluated and used, while also ensuring that such hurdles do not unnecessarily stall them from saving thousands of lives.

In the next section we will explore microorganisms that have been genetically engineered to be quite different from their normal brethren.

Further Reading

The following resources may be of interest for those wanting to learn more about genetically modified crops:

- Dominic W. S. Wong's book *The ABCs of Gene Cloning*
- Felicia Wu and William P. Butz's book *The Future of Genetically Modified Crops*
- The National Research Council's workshop summary "Genetically Engineered Organisms, Wildlife, and Habitat"
- C. Neal Stewart, Jr.'s book *Genetically Modified Planet*
- Martin Qaim's book *Benefits of Genetically Modified Crops for the Poor: Household Income, Nutrition, and Health*
- GMO Compass's webpage "Plants" at http://www.gmo-compass.org/eng/database/plants/

Genetically Modified World, Part II: Microorganisms

Microorganisms with amazing abilities

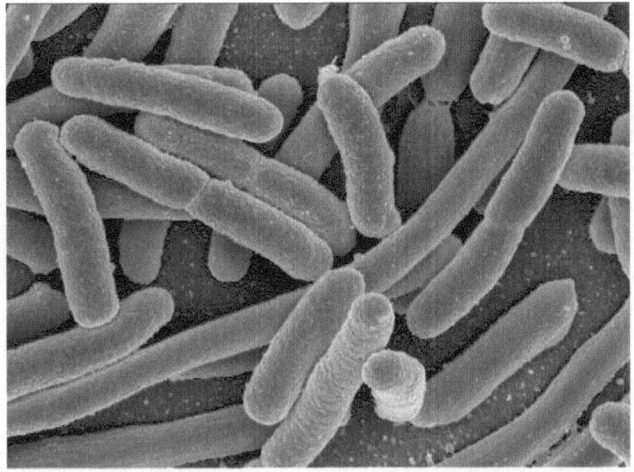

Decades of study into how to manufacture products using genetically modified bacteria (usually E. *coli*, shown above at high resolution), have made available to us today an astounding array of medical and food creations generated in such modified E. *coli*; we've even coaxed these critters into producing silk and pictures!

When thinking about genetically modified organisms (GMOs), the first things that usually come to mind are GM plants (which were discussed in the previous section) and animals, which overlooks possibly the most important group of all: GM microorganisms.

Through decades of development, the products of these miniscule critters have come to affect our daily lives. They have already become essential to the medical and food processing industries. Perhaps one day, they will also be critical to the agricultural and energy-production businesses, among others.

THE FIRST GMO

In the early 1970s, Herbert W. Boyer and colleagues (at the University of California, San Francisco, and Stanford University School of Medicine) created what is generally considered to be the first genetically modified organism. That organism was a microorganism, specifically the bacteria *Escherichia coli* (*E. coli*). (While we're most familiar with pathogenic *E. coli* strains through the headlines they occasionally make, there are actually many strains of *E. coli* that naturally live in our stomachs as a part of its rich, and essential, micro-flora.) Bacteria like *E. coli* have relatively simple genetics compared to plants and animals, which is part of why researchers have used them in genetic manipulation studies for decades. Additionally, *E. coli* reproduce quickly and can be grown in large numbers relatively cheaply.

So how did Boyer and colleagues "modify" the genetics of *E. coli*? Very simply put, the researchers put the DNA of a gene known to confer antibiotic resistance into the bacteria, and showed that the modified bacteria became resistant to the antibiotic. But, of course, it's a bit more complicated than this. The antibiotic resistance gene was not sent into the bacteria all alone; it was delivered in a package called a "plasmid," which is a piece of DNA (usually circular) that tells the cell to make more copies of the plasmid, along with any genes that are in it, like the antibiotic resistance gene. Boyer's team showed that two different plasmids could be cut open and recombined to make a new, unique plasmid (this is why it is called "recombinant" technology). This concept laid the foundation for a whole new enterprise.

What exactly were the implications of Boyer's work? That desired proteins could be made by bacteria in vast quantities, without the need for the harsh chemicals and extreme conditions that chemical synthesis often requires. On a single plasmid, researchers could not only put a gene that makes the bacteria become antibiotic resistance, but they could also insert a gene that would make the bacteria manufacture some protein of interest. While bacteria are pretty good at incorporating foreign DNA they're exposed to, compared to most organisms, it is still nowhere near 100% successful. Consequently, to make sure the bacteria have taken up the plasmid researchers want them to take up, they can expose the bacteria to an antibiotic. The ones with the plasmid will live, while the others will die. With the desired bacteria selected out, they can be easily, rapidly grown to great numbers and then stimulated (by exposure to a certain protein) to make the desired protein. Because the bacteria have been made to focus all their energies on producing this protein, they often will die as the protein is being harvested, but produce a lot of protein in the process. This strategy has become an essential technique, used routinely in laboratories across the world today for a variety of purposes.

MEDICAL APPLICATIONS

Many quickly realized the implications of this new recombinant technology. In the mid-1970s, Boyer went on to found Genentech, often considered the first biotechnology company. In 1977, the first human protein created in another organism was produced; Genentech made E. *coli* that could turn out the human protein somatostatin (which regulates hormones in the body, among other functions).

A year later, Genentech succeeded in modifying bacteria to make human insulin, which became the first GMO-produced product on the market in 1982. Insulin is needed to treat patients with type 1 diabetes, and can control the progression of type 2 diabetes in some patients. Insulin specifically stimulates the body to take sugar (specifically glucose) out of the blood stream and put it into different tissues (i.e. the liver, fat tissue, and muscle), where it is stored until needed. When insulin levels are low, or a person becomes insulin resistant (their tissues no longer respond to normal levels of insulin), then high blood sugar levels can result, which can cause diabetes. Before Genentech created a "recombinant" source of human insulin, all medical insulin was harvested from the pancreases of pigs or cows. In 2007, sales of insulin products estimated around $8 billion globally, and they will almost surely increase with time. (All GM-produced proteins must be chemically pure and safe, by the way, just as any non-GM protein must, to be approved for use.)

In the 1980s, Genentech competed with other biotechnology companies to be first in creating recombinant blood clotting factors. Why clotting factors? Clotting factors are necessary for people with hemophilia to survive. But in the early 1980s, significant amounts of stored clotting factors that had been derived from human plasma were found to be contaminated with HIV. Genentech won the race; its scientists were able to make several clotting factors in the laboratory (e.g. anti-hemophilic factor, or factor VII, and factor IX).

By now, a wide variety of other drugs, hormones, and other products have been created in microorganisms for the medical market. This includes erythropoietin (which regulates red blood cell production), interferons (which stimulate the immune system), vaccines (e.g. against the Hepatitis B virus), human growth hormone (hGH), and many more. Most recently in the medical field, bacteria are in clinical trials that may prevent cavities, and other bacteria are being created that may possibly block HIV infection.

IT'S WHAT'S FOR BREAKFAST!

Today, many of the food additives we consume on a daily basis are made by GM microorganisms. To name just a few, the list includes vitamins (B2, C), amino acids that improve flavors (e.g. aspartame), food additives (e.g. xanthan), food preservatives (e.g. nisin), enzymes used in food during specific chemical reactions (e.g. making cheese, breads, certain alcohols, and some sugars), and many more. While some of the microorganisms used may naturally make these products and are only coaxed into producing higher quantities, other microorganisms do not and must have the foreign gene inserted in them to do so.

RELEASE INTO THE ENVIRONMENT

The U.S. Environmental Protection Agency (EPA) approved the first use of genetically modified bacteria in the environment in 1985. In this particular case, researchers had taken bacteria that normally encourage ice formation on plants (specifically *Pseudonomas syringae* and *P. fluorescens*) and got rid of a gene that the bacteria needed to do this. Consequently, plants colonized with these modified bacteria do not form frost until around 23 degrees

Fahrenheit, saving them from damages brought upon plants by an early frost, or unusually cold weather.

Researchers soon explored another agricultural area that could greatly benefit from GM bacteria: the creation of improved nitrogen-fixing bacteria to help increase legume growth. Legumes (which include beans, lentils, peanuts, soy, alfalfa, clover, and more) are distinguished by their ability to capture nitrogen from the atmosphere, which they accomplish through a symbiotic relationship with a group of bacteria (rhizobia) that live in their roots. The bacteria *Sinorhizobium meliloti*, which can colonize multiple types of legumes, has been of particular interest. In some field studies, legumes with genetically modified *S. meliloti* had improved plant growth. However, some of these studies revealed an area of concern for using GM microorganisms: While the introduced bacteria (which is dependent on the legume hosts) was often undetectable after a few weeks or a year, sometimes it persisted in soil for more than two years. There have also been concerns that the genes of persistent GM microorganisms may be transferred to other, native bacteria, or to the organisms that consume the bacteria. Ongoing work is being done to address such issues.

More recently, genetically modified bacteria has been used to clean up the environment. In one such example, researchers have used rhizobia bacteria (specifically our friend *S. meliloti* again) and modified them to be able to break down a compound related to TNT. When alfalfa seeds were coated with these modified bacteria and planted in soil with the compound, the bacteria could reduce the pollution levels while stimulating plant growth.

IN THE WORKS...

A variety of creative applications for using genetically modified microorganisms have been explored over the last decade. In 2005, light-sensitive *E. coli* were created and used to create a kind of slow-action, high-resolution camera. An array of yeast strains have been modified that can convert sugars to ethanol (which we currently mostly harvest from corn), which could be a valuable alternative energy source. Similarly, many efforts have been invested in creating gasoline (crude oil) from microorganisms. In 2010, there was a breakthrough in generating stronger, lighter silk from bacteria. They've become more efficient than ever (generating one to two grams of silk per liter of bacteria), but how to efficiently put these silk proteins together into useable threads remains a big problem.

THE FUTURE

GM microorganisms have certainly been a boon for the medical and food industries, not to mention general laboratories everywhere. They also hold promise for creative applications and possibly for well-needed alternative energy sources. However, there is still concern over the release of GM microorganisms into the environment; in general, more observational studies on the effects of these microorganisms on their new habitats would be beneficial.

In the next and final section of this three-part series on the "Genetically Modified World," we will explore the amazing array of genetically modified animals that have been created.

Further Reading

The following resources may be of interest for those wanting to learn more about genetically modified microorganisms:

- Dominic W. S. Wong's book *The ABCs of Gene Cloning*
- The National Research Council's workshop summary "Genetically Engineered Organisms, Wildlife, and Habitat"
- Gary Walsh's article "Therapeutic insulins and their large-scale manufacture" at http://dx.doi.org/10.1007/s00253-004-1809-x
- Steven W. Pipe's article "Recombinant clotting factors" at http://dx.doi.org/10.1160/TH07-10-0593
- The United States Environmental Protection Agency (EPA)'s website at http://www.epa.gov/
- Mareike Viebahn et al.'s article "Effect of Genetically Modified Bacteria on Ecosystems and Their Potential Benefits for Bioremediation and Biocontrol of Plant Diseases – A Review" at http://dx.doi.org/10.1007/978-90-481-2716-0_4
- *National Geographic*'s article "Genetically Modified Bacteria Produce Living Photographs" by Joab Jackson at

http://news.nationalgeographic.com/news/2005/12/1206_051206_bacteria_photos.html
- *MIT Technology Review*'s article "Making Spider-Strength Materials" by Katherine Bourzac at http://www.technologyreview.com/biomedicine/25922/page1

Genetically Modified World, Part III: Animals

Beefed-up fish and drug-lactating livestock

Over the last two decades, cows have been genetically modified to not only grow faster, become larger, and be disease resistant, but also to produce a variety of pharmaceutical drugs in their milk.

For better or worse, we are constantly changing the world around us. The animals closest to us often experience this process the most. For thousands of years, people have domesticated animals for food, clothing, various other products, and companionship. With the advent of genetic modification (GM), we have become able to mold animals much more quickly and directly than ever before. While this technology has already proven to be quite useful and promises to be even more advantageous, it must be used with much caution and prudence.

Since their creation over three decades ago, genetically modified animals have quickly found their way into many different fields. From their inception in research laboratories, where they continue to be used to this day to help solve basic biology questions, the farm industry seized hold of GM animals and by the early 1990s many fast-growing, disease-resistant livestock animals had already been created. At the same time, the pharmaceutical industry realized the potential of GM technology and generated livestock that could produce all kinds of drugs, quickly and in large quantities, in farm animals' milk. Even the pet industry reached out for a slice of the GM pie, creating the "GloFish," zebrafish that fluoresce bright colors. Truly, GM applications to animals have come a long way in a very short time.

GM ANIMALS TO STUDY HUMAN DISEASES

The first animal to be genetically modified was the mouse, not surprising given its century-long relationship with scientists. Because mice are similar to humans in many ways, take up relatively little space, and reproduce quickly and in great numbers, they are ideal models for us to study when trying to better understand our own biology (as was explored earlier in this chapter).

It was nearly 40 years ago, in 1975, that Rudolf Jaenisch, Hung Fan, and Byron Croker (at The Salk Institute for Biological Sciences in San Diego and the Scripps Clinic in La Jolla) showed that a virus could be used to deliver genes to a mouse. Specifically, they harvested mouse embryos, infected them with the natural, unmodified Moloney murine leukemia virus, implanted these embryos into a surrogate mouse mother, and found the baby mice had DNA in their tissues from the virus. (As an interesting aside, it's been found that up to 8 percent of the human genome has DNA from viruses; a virus integrating their genes into other genomes is not a new phenomenon.)

Researchers quickly realized that they could use viruses to deliver all sorts of different genes into animals. By the early 1980s, several mouse lines (lineages descended from the original modified parents) had been made that had "foreign" genes inserted into their genomes and could pass these genes on to their offspring. (Such animals are also commonly called "transgenic.") While this was originally done using viruses, researchers found they could also inject DNA right into a fertilized mouse egg, although this approach is often less effective.

So what is the point of altering the genetics of mice? As mentioned above, mice are great models for studying human biology. Through genetic modification, mice have been made, for example, with development abnormalities that mimic ones seen in humans to help us better understand how they are caused and, consequently, how they can be treated and/or prevented. Today, mouse models of countless human diseases and conditions have been created, and more are made in laboratories all the time, assisted by the completion of the sequencing of the mouse genome in 2002.

Many other animals have also had their genetics altered to study different aspects of human biology. Researchers studying fruit flies (specifically *Drosophila melanogaster*) quickly took advantage of burgeoning GM technology. Fruit flies were the first widely used model organism. By the early 1980s researchers were adding foreign genes to fruit fly genomes to better understand development, which was already well characterized in them. A wide range of other animals have been genetically modified that may help us better understand ourselves. Most recently, in 2009, scientists were able to introduce a gene into a primate for the first time (in marmosets); these monkeys and their offspring glow green under UV light.

GENETICALLY MODIFIED FARM ANIMALS

With GM technology becoming well understood in mice, the livestock industry recognized this technology's great potential. By the early 1990s, researchers had made the first transgenic pigs, sheep, dairy cows, and fish, signaling the beginning of a GM farm animal revolution.

The genes most frequently inserted are used to make the animals produce more growth hormones, which consequently make them grow faster and larger, and often produce higher quality meat or milk. Farm animals have also been genetically modified to be more disease resistant; the first pigs resistant to viral infections were created in 1992, and in 2005 cows were made that were resistant to bacterial infection related to mastitis (inflammation of the breast tissue, a common ailment of dairy cows). Today, disease resistances have also been genetically added to poultry and fish, among others.

The fish industry is often overlooked when thinking of "farm" animals, but as an industry that processes 500 million pounds of meat annually, it has reason to explore GM technology. In 1992, Choy Hew and colleagues (at the Hospital for Sick Children and Departments of Clinical Biochemistry and Biochemistry at the University of Toronto) created GM Atlantic salmon that, by one year old, were an average of twice or six times as large as their non-GM relatives, with some fish as much as 13 times larger. Again, this was accomplished by injecting the gene of a growth hormone into the fish eggs.

Several other farm fish have been genetically modified, including trout (which also belongs to the salmon family), carp, and tilapia, usually to improve their growth rates. For example, in 2001, transgenic tilapia were created that could more efficiently digest dry protein diets, resulting in fish that were two to five times larger than the non-transgenic tilapia. Such fast-growing GM animals are often referred to as "feeding machines."

However, already researchers have discovered "natural" limits to the GM approach of increasing farm animal growth. Interestingly, in 2001 when a strain of domesticated, commercial, fast-growing rainbow-trout had growth hormone genes inserted into their genomes, their growth did not significantly increase, although their slower-growing, wild relatives had a 17-fold weight increase with the addition of the gene. The domesticated trout may have already reached a kind of growth ceiling due to decades of selection by fish farmers. Other studies have suggested that although GM fish may grow significantly faster, they may be more susceptible to disease and have reduced swimming abilities. Clearly there are currently limits to what GM additions can accomplish for the meat industries.

Some environmentally-conscious researchers have created animals that produce less waste. At the end of the 1990s, researchers at the University of Guelph in Ontario, Canada, created the "Enviropig," a pig that produces less phosphorous in its manure. Why is this good for the environment? Because the manure of pigs eating commercial diets contains high amounts of phosphorous, which can leach into the nearby watershed, contaminating ponds, lakes, and rivers, and result in algal growth, fish kills, and undrinkable water. The Enviropig produces a protein (introduced by scientists) that allows it to absorb more of the phosphorous in its diet, so 30 to 70 percent less phosphorous comes out the other end. In February 2010, the promising Enviropigs received approval by the Canadian government to live outside of research facilities, in facilities separated from other animals.

THE PERPETUAL FIGHT AGAINST PESTS

Researchers have furthermore used GM technology to defend crops from pests by targeting the pests themselves. The pink bollworm (*Pectinophora gossypiella*) is a very resilient pest of cotton farmers (it's even resistant to cotton genetically modified with pesticides, which was mentioned in the first section of the Genetically Modified World series). Consequently, researchers created genetically modified, sterile pink bollworms in attempts to keep their numbers down; sterile males released over the crops mate with females, but the eggs do not become fertilized, causing a massive population drop of the pests.

Researchers have even altered the bacteria in the guts of insects in hopes of getting rid of pests. The bacteria (*Xylella fastidiosa*), carried in the guts of a leafhopper insect, is responsible for Pierce's disease in grapevines. Instead of attacking the guilty bacteria directly, researchers tried to improve its competitor, another bacteria (*Alcaligenes*) that also resides in the leafhopper's guts. While it's hoped that the GM bacteria will displace the culpable bacteria in the guts, field trials are needed to confirm this. However, because any grapevines that the studies are conducted on will have to be destroyed afterwards (GM studies require very stringent safety protocols until they are confirmed to be safe), researchers are having a hard time finding an area to test these GM leafhoppers on.

The battle against pests is a tough one: Every 60 days a new disease or pest finds its way into California alone. GM technologies are a valuable addition to the suite of tools used to fight them.

GOT ANTI-THROMBIN?

One of the areas of GM technology that is most heavily investigated today is the use of animals as a means to produce pharmaceutical products or biomaterials. Since the early 1990s, livestock have been genetically modified to produce a wide range of pharmaceutical proteins or biomaterials in their milk, including: coagulation factors, anti-clotting agents, anti-thrombin, calcitonin, altered milk proteins, tissue plasminogen activator, vaccines, and even spider silk, among many more (including some that are also produced in bacteria, as was discussed in the previous section).

If some of these proteins were already made in bacteria, what's the advantage to making them in animals? One big reason is that some proteins cannot be properly created in bacteria and need a mammalian system to be produced correctly in an active form. Additionally, livestock can rapidly make large amounts of a desired protein: In one liter of milk, two grams of the introduced protein can be produced. Dairy goats were created in 1991 that could milk out anti-thrombin III (important for treating thrombosis and pulmonary embolism); the estimate was that 75 goats could produce 75 kilograms in one year, meeting the world demand with a small herd. However, there have occasionally been some problems in such GM livestock animals, ranging from sterility, to premature cessation of lactation, to other rare undesired side-effects.

In addition to farm animals, mice have also been used as "drug factories." Mice were created in 2002 that produced desired proteins, including erythropoietin, in their urine. However, while this means the protein can be harvested throughout the lifetime of the animal, it is probably not as efficient as milking livestock.

HURDLES

As with any developing technology, there are still many hurdles to overcome in the field of genetically modified animals. When genes are added to livestock to make them grow more efficiently or produce pharmaceutical drugs in their milk, there can be undesirable side-effects that are very difficult to predict. Additionally, although cows and goats can produce large quantities of pharmaceutical drugs in their milk, it can still be difficult to purify the drugs out of the milk, separating them from the normal milk proteins. There are also many concerns over how GM animals may affect their environment, which can only begin to be predicted through extensive laboratory tests where as many conditions as possible are tested. While GM technology holds immense promise, foresight and careful analysis is clearly required to properly fulfill it.

Further Reading

The following resources may be of interest for those wanting to learn more about genetically modified animals:

- Dominic W. S. Wong's book *The ABCs of Gene Cloning*
- The National Research Council's workshop summary "Genetically Engineered Organisms, Wildlife, and Habitat"
- Eduardo Melo et al.'s article "Animal Transgenesis: State of the Art and Applications" at http://www.webcitation.org/5k5aYeGAs
- Shao Jun Du et al.'s article "Growth Enhancement in Transgenic Atlantic Salmon by the Use of an 'All Fish' Chimeric Growth Hormone Gene Construct" at http://dx.doi.org/10.1038/nbt0292-176
- Robert H. Devlin et al.'s article "Growth of Domesticated Transgenic Fish" at http://dx.doi.org/10.1038/35057314
- The University of Guelph's webpage "Enviropig" at http://www.uoguelph.ca/enviropig/index.shtml
- Sabrina Richards's article "Will GM Insects Help Stop Disease?" in *The Scientist* magazine at http://www.the-scientist.com//?articles.view/articleNo/34005/title/Will-GM-Insects-Help-Stop-Disease-/

CHAPTER 4

BETTER UNDERSTANDING CANCER

Cancer Vaccines, Part I

Field is in the beginning stages

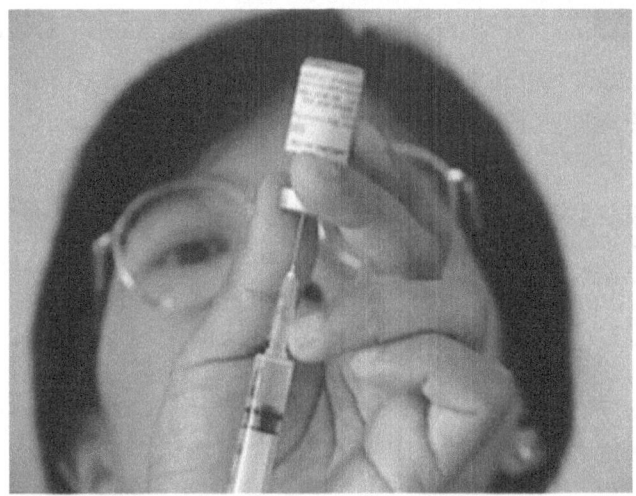

Vaccination against certain viruses can prevent most liver and cervical cancers.

Cancer treatments today involve an astounding array of drugs. Often these drugs, such as ones used in chemotherapy, have a wide variety of negative side-effects, including fatigue, pain, and hair loss, to name just a few. An emerging, appealing alternative is vaccines against cancer.

While the field is just in the beginning stages, cancer vaccines hold great potential as a possibly more effective, and less detrimental, approach to treating cancer by working with the body's own natural immune system.

In many ways, a cancer vaccine works very similarly to a normal flu vaccine. For both vaccine types, a "foreign" material enters a patient's body in order to create a strong immune response against the material. The vaccine itself contains some part of the foreign material that will let the body identify the invader, such as parts of proteins and cells, or even entire cells. When your body is exposed to anything foreign it wants to expel, there are generally three main steps that take place to cause a full-on assault against the invader:

First, the body must identify the material as being "foreign," an invader in the body. T cells, a type of immune system cell, accomplish this.

Second, the body must generate an immune response against the invader, which is in the form of antibodies (produced by B cells).

Third, rallied by the T cells, the body must strengthen the immune response against the invader.

PREVENTIVE VS. THERAPEUTIC

In the field of cancer vaccines, there are two main groups of vaccines: "preventive" vaccines and "therapeutic" vaccines. While cancer is not what is typically thought of as an "infection," evidence has shown that some cancers can be caused by infectious viruses. And so, preventive cancer vaccines have focused on vaccinating people against these potentially cancer-causing viruses. Again like the usual flu vaccine, preventive cancer vaccines are given to patients before they are infected, the goal being to prepare the immune system for a real, future attack. Remembering its encounter with the vaccine, the immune system should have a head start on producing antibodies for any real encounters with the virus in the future, preventing infection from even getting a foothold.

Therapeutic cancer vaccines, on the other hand, are for patients who already have cancer. Their aim is to prevent the progression of the cancer (e.g. reduce tumor size and the spread of the cancer) by stimulating an immune response against it. While some therapeutic cancer vaccines have had promising results, preventive cancer vaccines have been more successful.

THERAPEUTIC CANCER VACCINES

The rest of this chapter will focus on preventive cancer vaccines, while therapeutic vaccines will be explored in the following chapter. At the time of this writing, preventive vaccines against two potentially cancer-causing viruses, hepatitis B and the human papillomavirus, are the only cancer vaccines approved by the FDA.

HEPATITIS B AND LIVER CANCER

Hepatitis B (HBV) is a virus that causes about 80% of all liver (hepatic) cancers. (The other 20% are thought to be caused by alcohol consumption, obesity, and other conditions.) While liver cancer is around the fifth most common cancer globally, it comes in third for its resultant fatalities.

HBV can be easily transmitted through sexual contact or other exchanges of blood or fluids, such as through dirty needles or childbirth. The HBV virus attacks the liver specifically, causing it to become inflamed, although a person with HBV may not show any symptoms outwardly. Chronic HBV infection can lead to liver cancer. Just like the flu, there are different strains of HBV, and recent research has shown that some strains may be more likely to cause liver cancer than others. Fortunately, there are vaccines for HBV that can prevent viral infection, and consequently the development of liver cancer from it.

Effective HBV vaccines were approved by the FDA in 1981. The virus had been suspected for decades before this to cause liver cancer. The HBV vaccine was the first FDA-approved cancer vaccine. Today, there are two main HBV vaccines: Recombivax HB, made by Merck, and Engerix-B, created by GlaxoSmithKline. Both HBV vaccines contain part of a protein found on the surface of HBV, stimulating the body to build up an immune response against this piece of protein to prevent infection with the virus. (Rest assured, it is impossible to contract HBV from vaccines like this.) HBV vaccines are now quite established and most children in the U.S. are vaccinated at birth. Great efforts have been made to establish at-birth HBV vaccination in areas where HBV is particularly prevalent, such as sub-Saharan Africa or Eastern Asia where over 80% of the global HBV infections occur.

HUMAN PAPILLOMAVIRUS AND CERVICAL CANCER

Fifteen years after the FDA approved the HBV vaccine, the second, and currently only other, cancer vaccine was approved. It was for the human papillomavirus (HPV), which can cause cervical cancer. (For more information on this vaccine, see http://www.cancer.gov/cancertopics/factsheet/Risk/HPV.) Every year, HPV infects around half a million women and kills around a quarter of a million women globally (including 4,000 deaths in the U.S.). It is transmitted through sexual contact and is extremely prevalent; it's estimated that around half of all sexually active men and women get HPV at some point in their lives, and about 20 million Americans currently have HPV, though most may never know it. Like HBV in the liver, a persistent cervical infection of HPV, which may or may not show clear symptoms, can lead to cervical cancer.

In June 2006, the FDA approved HPV vaccines for girls and women ages nine to 26. The two HPV vaccines (Gardasil and Cervarix, again made by Merck and GlaxoSmithKline, respectively) very effectively protect against different HPV strains that cause cervical cancer. Early trials showed it to be nearly 100% effective in young women. Like the HBV vaccines, the HPV vaccines contain parts of proteins found on the surface of the virus, stimulating the body to recognize and attack the virus itself if it should encounter HPV. It is recommended to administer the vaccines to girls between the ages of 11 and 12, long before any potential exposure, but women up to 26 may also benefit from vaccination.

OTHER VIRUSES THAT CAUSE CANCER

There are several other types of cancers known or suspected to be caused by viruses, though no vaccines currently exist for these viruses. For example, strains of the Epstein-Barr virus (EBV) (which causes mononucleosis, commonly known as "mono") are linked with several types of cancers: Burkitt's lymphoma, gastroadenocarcinoma, and nasopharyngeal carcinoma. EBV is thought to infect over 95% of people at some point in their lives. No vaccines are currently available for EBV, but some are in clinical trials and may one day prove an effective defense against not only EBV, but these cancers as well.

Although the idea of having a vaccine against cancer seems like science fiction, these vaccines are already here and, in the case of HBV, have been here for decades. However, these treatments currently only apply to the limited group of cancers that are caused by viruses. It has proven much more difficult to create successful therapeutic cancer vaccines that help a patient who already has cancer. But despite great difficulties, therapeutic cancer vaccines have already made it to clinical trials for a variety of cancers, including lung, breast, prostate, colorectal, melanoma, lymphoma, and kidney cancer. Some of these vaccines have had very promising results, and these will be the focus for the next section, Cancer Vaccines, Part II.

Further Reading

The following resources may be of interest for those wanting to learn more about cancer vaccines.

- The National Cancer Institute at the National Institutes of Health's website at http://www.cancer.gov/
- The National Cancer Institute at the National Institutes of Health's webpages "Cancer Vaccines" at http://www.cancer.gov/cancertopics/factsheet/cancervaccine and "Human Papillomavirus (HPV) Vaccines" at http://www.cancer.gov/cancertopics/factsheet/prevention/HPV-vaccine
- Shijian Liu et al.'s article "Associations Between Hepatitis B Virus Mutations and the Risk of Hepatocellular Carcinoma: A Meta-Analysis" at http://dx.doi.org/10.1093/jnci/djp180
- E. Giarelli's article "Cancer Vaccines: a new frontier in prevention and treatment" at: http://www.ncbi.nlm.nih.gov/pubmed/18154203

Cancer Vaccines, Part II

Therapeutic cancer vaccines

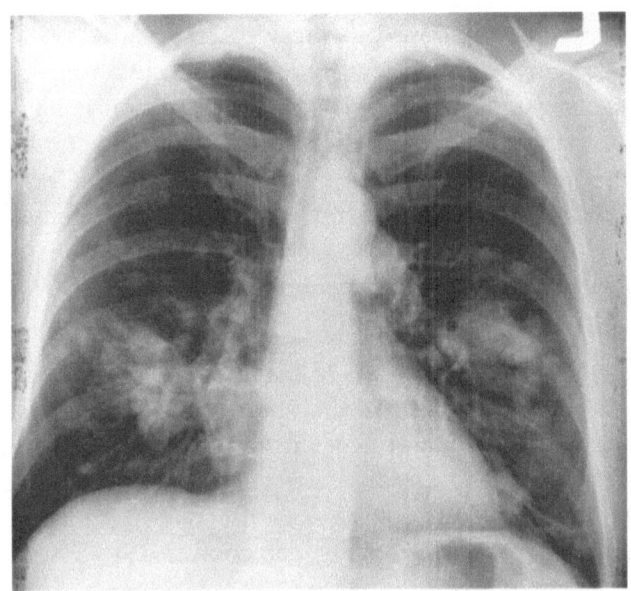

While vaccines have been developed to fight many types of cancers, one such promising vaccine is Lucanix, which is being tested in clinical trials for treating lung cancer.

Nearly 1.5 million people in the United States were estimated to have been diagnosed with cancer in 2009, according to the American Cancer Society. Although an amazing variety of drugs have been developed to fight cancer, it remains the second leading cause of death in this country (the first is heart disease), claiming the lives of just over half a million Americans last year. It is responsible for nearly one in four deaths in America.

Clearly, better approaches are needed to effectively fight cancer. An emerging alternative to traditional drugs is cancer vaccines, which function by stimulating the patient's own immune system to fight the cancer. The first therapeutic cancer vaccine to be approved by the FDA, a vaccine called Provenge which is used to treat prostate cancer, received its approval in April of 2010, although many other cancer vaccines also have promising results from clinical trials.

Just like flu vaccines, cancer vaccines are intended to stimulate an immune response from the patient to reject a "foreign" material. There are two main groups of cancer vaccines based on what they target: "preventive" and "therapeutic." Preventive cancer vaccines immunize against viruses that cause cancers, such as hepatitis B and the human papillomavirus (the viruses that cause liver and cervical cancer, respectively). There are successful, FDA-approved preventive cancer vaccines for these viruses, which were discussed in the previous chapter. In this section the second group of cancer vaccines will be explored: therapeutic cancer vaccines.

Therapeutic cancer vaccines are designed to help patients who already have cancer. The goal is to make the body have an immune response against the cancer, so that the body can effectively fight it. Because cancer is made up of cells from the patient, it is not truly a "foreign" material and so it is normally difficult for the immune system to identify it as a threat. Consequently, cancer can hide from the immune system, avoiding detection and attack.

The key principle of cancer vaccines is that the immune system needs help, or stimulation, to recognize the cancer as a threat and attack it. Cancer vaccines usually contain some component of the cancer, either parts of proteins and cells or entire cells, to give the immune system a clear target to attack. However, researchers have to be particularly careful about what they make a cancer vaccine out of: If a protein is made by many cells other than the tumor cells, the normal cells may be accidentally targeted.

This has been greatly overcome by identifying proteins that only tumors make in large amounts. Although it is much harder to develop therapeutic cancer vaccines than preventive ones because of these complications, therapeutic cancer vaccines are being tested in clinical trials for treating lung, colorectal, breast, prostate, melanoma, lymphoma, and kidney cancers, and many have yielded promising results and may be nearing FDA-approval. So, what cancers are these vaccines targeting?

LUNG CANCER

The most common malignant cancer is lung cancer, killing more people each year than any other cancer: the National Cancer Institute of the U.S. National Institutes of Health estimates that in 2013 lung cancer will kill nearly 160,000 people in the United States (while nearly 230,000 new cases will be diagnosed), and these estimates are similar to what they have been for the past several years. (For details, see

http://www.cancer.gov/cancertopics/commoncancers.) A therapeutic cancer vaccine for non-small cell lung cancer (NSCLC) (which makes up the majority of lung cancer cases) is in a phase III clinical trial. The vaccine, called Lucanix (Belagenpumatucel-L), is produced by a San Diego-based biopharmaceutical company, NovaRx Corporation.

Lucanix is intended to fight NSCLC by interfering with a protein called transforming growth factor-beta (TGF-beta). TGF-beta is often elevated in patients with NSCLC, which causes the immune system to become suppressed and not attack the cancer. Lucanix is made up of four types of lung cancer cells that have been genetically modified to inhibit the secretion of TGF-beta, ideally increasing the ability of the patient's body to identify the cancer as an "invader" and fight it off. In phase II clinical trials, patients with advanced NSCLC lived an average of 16 months longer with Lucanix. The NovaRx Corporation has become the first to demonstrate that inhibiting TGF-beta can increase the body's immune response to a tumor, which may result in longer survival. This approach could potentially be applied to other cancers. Although no unusual negative side effects have been reported, caution must be used; TGF-beta has other key roles in the body (such as preventing cells from growing out of control), and care must be taken to ensure that other functions are not undesirably disrupted.

As of 2012, Lucanix is being used in an ongoing, phase III clinical trial for treating non-small cell lung cancer. (For details on these trials, see http://clinicaltrials.gov/ct2/show/NCT00676507.) Near the end of my time as a graduate student at the University of California, Santa Barbara, a center in the area, specifically the Cancer Center of Santa Barbara, under the direction of Dr. Fred

Kass, became one of nearly 60 centers around the world to be involved in the phase III clinical trial. While Lucanix is not a cancer cure, it may be a relatively significant step in the right direction for improving the lives of patients with lung cancer. If successful, this clinical trial may be one of the last steps before FDA-approval.

COLORECTAL CANCER

Colorectal cancer is the second leading cause of cancer death in the United States, causing nearly 50,000 deaths in 2009, while around 147,000 people in the country were newly diagnosed. An appealing target for these cancer treatments is a protein called 5T4. 5T4 is made in high amounts by many different types of cancers (colorectal, ovarian, gastric, and kidney) but only rarely in other cells, and is believed to be involved in cancer metastasis. Oxford BioMedica, a United Kingdom-based biopharmaceutical company, developed a cancer vaccine to target 5T4 called TroVax. TroVax delivers the blueprints for 5T4 in a modified (benign) virus, just like a typical vaccine, the aim being to stimulate an immune response against the 5T4 protein.

In phase II clinical trials, TroVax was tested in more than 200 patients and 95 percent showed "potent and sustained immune responses" with minimal side effects. (To read more about these trials, see http://www.ncbi.nlm.nih.gov/pubmed/17727334.) A phase III clinical trial started in 2008 and underwent a series of amendments recommended by the FDA. In 2010 and 2011 results were published suggesting that enhanced survival of patients treated with TroVax correlated with specific blood parameters. (To read more about the latest reported findings, see http://dx.doi.org/10.1007/s00262-011-0993-7). Clinical trials have also been testing the efficacy of TroVax to treat kidney and prostate cancers as well.

BREAST CANCER

An estimated 40,000 women died from breast cancer in 2009. Mucin 1, or MUC1, is a protein made in high amounts by many different types of cancers: aside from breast cancer, it is also in prostate, lung, and colorectal cancers. Merck and Oncothyreon (a

German pharmaceutical company and a Seattle-based biotechnology company, respectively) created a potential cancer vaccine called Stimuvax to target MUC1. Stimuvax contains a short, synthetic protein that ideally generates an immune system response specifically against MUC1, consequently destroying the cancer cells that make it. After promising results in early clinical trials, in 2009 phase III clinical trials started with Stimuvax for breast cancer as well as lung cancer. However, in March of 2010 it was announced that further recruitment of participants was halted after one patient from a phase II trial contracted encephalitis (a serious condition caused by inflammation of the brain). Then in 2012 it was quite unfortunately announced that Stimuvax did not show an improvement in the overall survival of patients in general, although there may have been some effects observed for certain subgroups of patients, which has made some advocates hold out hope that the drug may be pursued for specific patient subgroups.

Additional breast cancer vaccines are being developed against other probable targets, such as human epidermal growth factor receptor 2, or HER-2. This growth factor, which is produced in abnormally high levels in 20-25 percent of breast cancers, is already a target of breast cancer drugs Trastuzumab and Lapatinib.

PROSTATE CANCER

Prostate cancer claimed the lives of around 27,000 Americans in 2009. The Dendreon Corporation, a Seattle-based biotechnology company, produced a prostate cancer vaccine called Provenge (sipuleucel-T), which, in April of 2010, became the first FDA-approved therapeutic cancer vaccine. Provenge aims to trigger an immune response against the cancer cells by zeroing-in on a protein made in large quantities by prostate cancers, called prostatic-acid phosphatase (PAP). The vaccine contains a patient's own blood cells that have been modified to make PAP. Clinical trials showed an average four months' increase in survival time, significantly better than current alternative prostate cancer drugs. As the first FDA-approved therapeutic cancer vaccine, Provenge holds much promise for treating patients with prostate cancer.

AND MORE...

Other cancers, such as non-hodgkin lymphoma, kidney cancer, and melanoma, also have different vaccines that have shown some success in clinical trials. A phase III clinical trial for treating follicular lymphoma (a type of non-Hodgkin's lymphoma) with the cancer vaccine BiovaxID showed remission time to be increased by just over a year. (To watch a video about this trial, see http://meetinglibrary.asco.org/content/36467, and to read about it see http://meetinglibrary.asco.org/content/33572-65.) A cancer vaccine called Oncophage, made by the New York-based company Antigenics, has been approved in Russia for treating kidney cancer. Trials showed it to increase survival by a year in some patients, although for FDA approval in the U.S. new trials are required. Cancer vaccines for melanoma have also met with success; BioVex Inc. created the vaccine OncoVEX GM-CSF (now known as talimogene laherparepvec). This vaccine contains a virus that only replicates in the melanoma tumor and creates proteins to stimulate the immune response there, enhancing rejection of the tumor. After BioVex was acquired by Amgen, in early 2013 it was announced that the vaccine was shown to shrink tumors in patients with late stage melanoma in a phase III clinical trial. (To read more about this vaccine, see

http://wwwext.amgen.com/media/media_pr_detail.jsp?year=2013&releaseID=1798143.)

With so many different therapeutic cancer vaccines in clinical trials and many showing very encouraging results, the outlook is bright for these novel cancer treatments. Others may soon be following the way of Provenge in receiving FDA-approval and become important weapons in doctors' ever-expanding arsenal to better fight cancer.

Next up: The concluding section of the cancer vaccine series will discuss using embryonic tissues and cells to vaccinate against cancer.

Further Reading

The following resources may be of interest for those wanting to learn more about cancer vaccines.

- The National Cancer Institute at the National Institutes of Health's website at http://www.cancer.gov/
- The National Cancer Institute at the National Institutes of Health's webpages "Cancer Vaccines" at http://www.cancer.gov/cancertopics/factsheet/cancervaccine and "Common Cancer Types" at http://www.cancer.gov/cancertopics/commoncancers
- E. Giarelli's article "Cancer vaccines: a new frontier in prevention and treatment" at http://www.ncbi.nlm.nih.gov/pubmed/18154203

Cancer Vaccines, Part III

Vaccinating with embryonic tissues

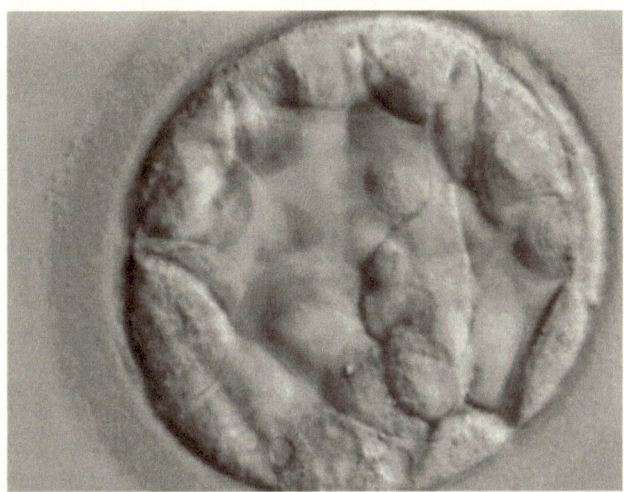

Vaccination with human embryonic stem cells (which can be harvested from early stage embryos, as shown here) has been recently shown to impart some immunity against cancer.

In the last two sections, we have covered a variety of cancer vaccine technologies, many of which are in clinical trials and some of which have received FDA-approval. These contemporary vaccines had their roots, naturally, in the laboratory.

In this section, the last of the three-part series on cancer vaccines, we'll return to the research laboratory to explore a preliminary cancer vaccine technology: using cells from embryos to help fight cancer. Although this style of vaccine will probably not be entering clinical trials any time soon, the biological lessons learned from these studies are very relevant to our understanding of cancer, and ultimately of how to better fight it.

In 2009, a research article was published by professors Yi Li, Zihai Li, and colleagues (at the University of Connecticut School of Medicine) showing that mice with colon cancer could be "vaccinated" with human embryonic stem cells. Vaccines contain an identifying piece of a potential biological invader (such as a protein only made by a certain virus). When the body is vaccinated, the immune system prepares itself to fight off the invader when it is encountered again. The idea of stimulating the immune system to fight off cancer is steadily gaining support.

But since cancer is threatening to the host, why do our immune systems not attack it in the first place? This problem is called "immune tolerance." The immune system usually fails to detect and attack cancerous tumors, and consequently many cancer treatments are currently being developed that stimulate the immune system to fight back (e.g. the growing field of cancer vaccines).

An intriguing insight into this state of immune tolerance can be found by looking at what happens to the immune system during pregnancy. It's been found that the body's immune response to a tumor is very similar to its response to a developing embryo. This helps explain why the 2009 study above, where mice were vaccinated using human embryonic stem cells, succeeded in treating cancer to some extent, but this is just part of a centuries-old story. The long history of studies illustrate three key points overall: (1) tumors and embryonic tissues have many similar molecules on their cells that are recognized by the immune system, and consequently antibodies made to identify and attack tumors can also do the same to embryonic tissues, and vice versa; (2) pregnancy, amazingly, confers some immunity against cancer; and (3) similar to pregnancy, an immune response against cancer can be generated by vaccinating animals with embryonic tissues.

TUMORS AND EMBRYONIC TISSUES ARE VIEWED SIMILARLY BY THE IMMUNE SYSTEM

The first published suggestion that tumors may have an embryonic nature dates all the way back to the late 1830s. It was proposed that tumors might be tissues that had become embryonic-like again. This idea was found later to be fairly accurate: It is now generally accepted that, in terms of the proteins they make, tumors are indeed more "embryonic" than the tissues they are derived from. By the late 1800s, researchers realized that they could better understand and combat cancerous tumors by better understanding normal development.

170

The focus of these studies shifted in the 1880s because the field of immunology was born (from research conducted by Louis Pasteur and Robert Koch) and many researchers started investigating creating vaccines to cure diseases. Cancer was no exception. As the field of immunology blossomed, in the 1920s and 1930s researchers found that tumors and embryonic tissues make similar molecules recognized by the immune system, molecules not found in non-cancerous adult tissues. (Molecules recognized by the immune system are called "antigens.") Antibodies can rally the immune system to fight off "invaders" identified by their antigens. So, in the 1960s and 1970s, it was found that antibodies made to specifically recognize animal tumors could also recognize embryos, and vice versa. These studies led to the idea of "oncofetal antigens," antigens that are made by both tumors and fetal/embryonic tissues.

PREGNANCY CONFERS SOME IMMUNITY AGAINST CANCER

Pregnancy is a fascinating biological condition. During the early stages of pregnancy, the immune system in the mother to-be is actually put on alert against the developing embryo. Specifically, in the 1960s a study reported that about 80% of blood serum (which normally contains antibodies) from pregnant women in the first two trimesters contained antibodies that reacted against embryonic tissue. Why more early pregnancies do not spontaneously abort due to this obvious immune response is still not well understood, but the placenta serves an important role in creating a protective interface between mother and child. However, given the above findings, it may not be too surprising to hear that these antibodies also recognized tumor tissues (but, importantly, not other adult tissues).

As the concept of oncofetal antigens developed, researchers explored a hypothesis with very practical applications: Because pregnant animals develop antibodies that can recognize tumor tissues, pregnancy may partially "vaccinate" an animal against cancer. Early observations in support of this theory are nearly 300 years old, as celibate nuns were thought to have increased incidences of breast, uterine, and ovarian cancer. Some researchers believed that this increased risk might be due to a lack of antibodies generated during pregnancy against embryonic-like tissues. Multiple studies in the 1960s through 1980s found support for this theory, repeatedly showing that women and animals that were multiparous (pregnant multiple times) not only had fewer spontaneous cancers, but multiparous animals were also more resistant to induced tumors (e.g. using known carcinogens in animal studies).

This phenomenon also seems to work the other way around: Correlations have been shown to exist between women with high amounts of antibodies that recognize oncofetal antigens, and low pregnancy rates. Because the antibodies also recognize embryonic tissues (as well as cancer tissues), some women who have had multiple miscarriages (spontaneous abortions) have an (unusually) high amount of these antibodies in their body (presumably all of the time).

EMBRYONIC TISSUES CONFER SOME CANCER IMMUNITY

As researchers found that pregnant animals have some immunity against cancer, the question was raised: Could injecting, or "vaccinating," with embryonic tissues alone similarly trigger an immune response and confer cancer immunity? In the late 1960s and early 1970s, many reports were published supporting this theory: animals vaccinated with embryonic tissues created antibodies that not only recognized tumors, but could even sometimes prevent the formation of tumors in the animals.

One such astonishing study was published in 1970, when researchers Elliott H. Stonehill and Aaron Bendich (at the Sloan-Kettering Institute for Cancer Research in New York) reported that rabbits immunized with nine-day-old mouse embryos created antibodies that recognized 72 different mouse tumors from 12 different tissue types. The antibodies, as expected, also reacted against embryonic tissues. Similar results were seen with human embryos.

These studies strongly suggest that vaccination with embryonic tissues not only triggers an immune response against cancer, but may also prevent it to some degree.

OVERCOMING THE PROBLEM OF IMMUNE TOLERANCE

But in order to truly harness the potential of using embryonic tissues to immunize against cancer, the problem of immune suppression must be better understood. In pregnant women and in patients with cancer, immune tolerance happens in an apparently similar manner. Its occurrence is both transient, and poorly understood.

In the 1960s, it was reported that while patients with sarcomas usually did not have detectable antibodies against the tumor, after the tumor was surgically removed, anti-tumor antibodies increased! This was not seen in patients whose sarcomas were not successfully removed. Similarly, antibodies against oncofetal antigens are present in pregnant women in the first two trimesters, then quickly disappear during the final trimester, and then reappear after birth.

The cause for the transient immune tolerance in both cases is unclear, though "blocking factors" in blood serum may facilitate this process; they were identified in the early 1970s. These factors most likely (1) prevent recognition of the embryo/cancer and/or (2) suppress the immune system in general (e.g. may include the widely studied factor TGF-beta). Clearly, to better understand immune tolerance in cancer, it would be helpful to better understand immune tolerance during pregnancy. There are probably many shared reasons for why pregnancies and tumors are not rejected by the body more often than they are.

While these studies showed many promising findings for fighting cancer, published reports significantly decreased after the mid-1970s. This was possibly due to decreased funding, and the infeasibility of progressing the studies further because of the dubious ethical nature of such human research. However, the connection between cancer and embryonic tissues/cells has been revisited recently using human embryonic stem cells (hESCs). A stable embryonic cell line had not been available to test in the previous studies. In 2009, professors Yi Li, Zihai Li, and colleagues (at the University of Connecticut School of Medicine) reported that vaccinating mice with hESCs resulted in an immune response against colon cancer. The hESCs conferred consistent, cellular and humoral responses against the colon cancer and significantly reduced the tumor size. This report, understandably, is bringing attention back to the longstanding field of cancer vaccination using embryonic tissues and cells. Other groups may also revisit this potentially promising tool for fighting cancer in the burgeoning field of cancer vaccines.

Further Reading

The following resources may be of interest for those wanting to learn more about cancer vaccines.

- Teisha J. Rowland's article "Cancer Vaccines: Using Embryonic Tissues and Stem Cells to Vaccinate Against Cancer" at http://www.allthingsstemcell.com/2010/05/cancer-vaccines/
- Bradley G. Brewer et al.'s article "Embryonic Vaccines Against Cancer: An Early History" at http://dx.doi.org/10.1016/j.yexmp.2008.12.002
- Yi Li et al.'s article "Vaccination with Human Pluripotent Stem Cells Generates a Broad Spectrum of Immunological and Clinical Responses Against Colon Cancer" at http://dx.doi.org/10.1002/stem.234
- Thomas P. Zwaka's commentary article "Stem Cell Vaccination Against Cancer: Fighting Fire with Fire?" at http://dx.doi.org/10.1038/mt.2009.287
- The National Cancer Institute at the National Institutes of Health's website at http://www.cancer.gov/ and webpage "Cancer Vaccines" at http://www.cancer.gov/cancertopics/factsheet/cancervaccine

A Childhood Cancer is Different from Adult Cancers

Far fewer mutations

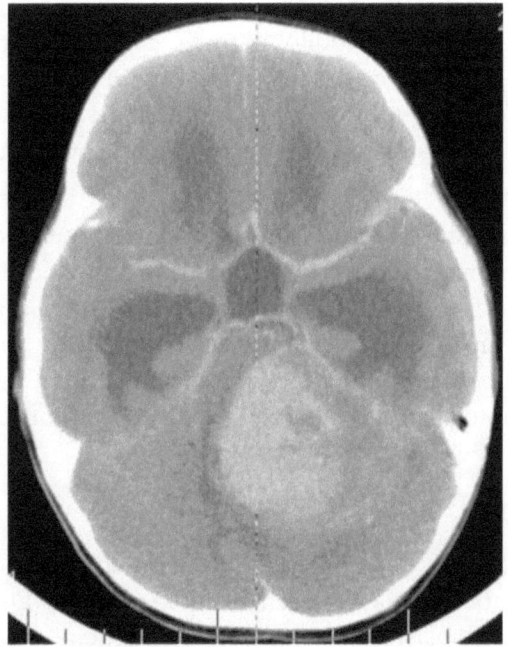

Researchers have recently found that childhood medulloblastomas (shown here) have many fewer mutations than different adult cancers.

In December of 2010, a study was published reporting that a cancer in children is very different from cancers in adults. Specifically, the study suggests that there are far fewer mutations present in some childhood cancerous tumors than in tumors of adult cancers. This could have significant implications for not only how we approach and treat childhood cancers, but how we combat cancers in general.

The study, which was published in the journal *Science*, was a large collaborative effort involving twenty different medical institutes, cancer centers, and universities in the U.S. that looked at the genetics of a very aggressive brain cancer found primarily in children called medulloblastoma. To read the actual study, see http://dx.doi.org/10.1126/science.1198056. Although medulloblastoma tumors originate in a certain area of the brain, the cancer can spread through the spinal fluid to other parts of the central nervous system (which includes the brain and spinal cord). Every year, about 1 in 200,000 children under the age of 15 are diagnosed with having medulloblastoma, and although therapies to treat them have improved, still many cannot be cured and often there are severe side-effects from the therapies. It's a devastating cancer. Medulloblastoma is very rare in people over 40, which most likely has to do with how medulloblastoma comes to be.

Although its identity has been hotly debated in recent years, it's now thought that medulloblastoma is created from cells that were primitive neural stem cells. Such neural stem cells are normally present during fetal development, but in patients with medulloblastoma, these stem cells have mutations in their DNA that led to the formation of a highly malignant tumor.

THE STUDY

To test whether, and how, medulloblastoma in children is different from adult cancers, the researchers looked at the DNA from 22 different medulloblastoma samples from children as well as DNA from several different adult cancers (including pancreatic, glioblastoma, colorectal, and breast cancer tumors). The DNA sequences of all known genes were analyzed in both tumors and in normal, noncancerous tissues taken from the same patients. In this way, the scientists could determine which genes were different, or mutated, in the tumors compared to healthy tissue. This is commonly done when studying cancer, and identifying which genes are mutated helps us better understand what causes cancer.

FAR FEWER MUTATIONS IN A CHILDHOOD CANCER THAN ADULT CANCERS

Interestingly, the scientists found that the medulloblastoma tumors in children had about five to 10 times fewer mutated genes than several different adult tumors studied (which were not medulloblastomas). However, while the childhood tumors had many fewer mutations, over one third were mutations known to disrupt gene function, which is a much higher percentage than in the adult tumors analyzed. This may simply be because the adult cancers are in older tissues, and these tissues have had a longer opportunity to gain random mutations over time, but it's difficult to tell. Either way, this shows that childhood cancer is very different from adult cancers.

THE LIKELY SUSPECTS

How can we use our knowledge of these differences between childhood and adult cancers to help us better combat cancer? To answer this question, it's important to look at exactly what the researchers found, which comes down to some very specific genes that probably play central roles in causing cancer.

Some of the genes commonly mutated in the medulloblastoma tumors belong to two different groups of well-studied genes, which are the Hedgehog and Wnt gene groups. (Many genes have somewhat whimsical names, mostly due to early studies in fruit flies; the Hedgehog genes, for example, don't have anything to do with the hedgehog animal.) These groups of genes are thought to primarily function during embryonic development and during some cancers. Specifically, the mutation of some Hedgehog and Wnt genes has previously been shown to correlate with the occurrence of different cancers, including medulloblastoma.

This recent study not only confirmed previous studies linking Hedgehog and Wnt genes to the occurrence of medulloblastoma, but the researchers also discovered that other genes, including ones not previously suspected, may also be involved in causing medulloblastoma. These newly suspected genes belong to the histone-lysine N-methyltransferase gene family, whose members actually control whether many other genes are active or not. Consequently, by having mutations in these identified genes (which are specifically the histone-lysine N-methyltransferase genes *MLL2* and *MLL3*), the tumors have changed the function of many other genes, causing a kind of snowball effect.

The researchers found that *MLL2* or *MLL3* were mutated, and consequently were no longer active, in 16 percent of the medulloblastoma tumors they looked at. This suggests that *MLL2* and *MLL3* normally play important roles in preventing the occurrence of medulloblastoma, that they are what are called "tumor suppressor genes." Additionally, it's known that these genes are also very important during normal brain development, which again emphasizes how this childhood cancer is different from adult cancers; the childhood cancer is almost certainly the result of something going wrong during development.

CONTRIBUTING TO HOW WE VIEW CANCER

While the most significant finding in this study was how many fewer mutations are present in a primarily childhood cancer, medulloblastoma, compared to adult cancers, understanding the specific genes revealed to be key to causing medulloblastoma will not only change how we view and treat childhood cancer, but will also help us better understand which genes are key players in causing cancer in general. It's becoming clearer that childhood cancer, shown here with medulloblastoma, is due to problems that manifest during fetal development.

Additionally, the family of genes that *MLL2* and *MLL3* belong to has already been suspected of being involved in the occurrence of many other types of cancer, and this most recent finding with medulloblastoma warrants further investigation of this gene family and its connections to cancer.

Lastly, because this childhood cancer has many fewer mutated genes than adult cancers, it gives researchers a few genes to really focus on; the genes mutated in medulloblastomas may be the few, most essential genes responsible for this kind of cancer and most likely other cancers. Applying these findings to other cancers may be difficult. In this study, childhood medulloblastomas were not compared to adult medulloblastomas (which are very rare, but may reveal additional differences), and clearly this study has shown that not all cancers are alike.

FIGHTING CANCER WITH GENE THERAPY

To date, over 80 cancers have had all of their known genes analyzed for apparent mutations. As we better understand which genes are most important for causing, and preventing, cancer, researchers can not only develop better methods for identifying the type of cancer a patient has, but can also develop better therapies to target these specific genes. For example, in medulloblastoma, thanks to this recent study, the development of drugs that target the genes *MLL2* and *MLL3* is a very appealing prospect, but a better understanding of these genes (and improvements in gene therapies in general) is needed before such therapies can become a reality.

Further Reading

The following resources may be of interest for those wanting to learn more about childhood and adult cancers.

- Teisha J. Rowland's blog "All Things Stem Cell" at http://www.AllThingsStemCell.com
- D. W. Parsons et al.'s article "The genetic landscape of the childhood cancer medulloblastoma" at http://dx.doi.org/10.1126/science.1198056
- *The Scientist*'s article "Fewer mutations in kids' cancer" by Megan Scudellari at http://www.the-scientist.com/news/display/57866/

Telomerase: The Fountain of Youth?

Enzyme can reverse tissue aging

Researchers at the Harvard Medical School recently found that an enzyme called telomerase can reverse tissue degeneration in mice. This may help us better understand and "treat" aging.

In November of 2010, an amazing and promising discovery was made: an enzyme called telomerase could, in mice, reverse tissue degeneration, one of the critical features of aging.

Telomerase is actually involved in many important biological functions, from cancer formation to the aging of our cells. While it was long thought that telomerase is involved in controlling the age of our cells, this study reported that it might also play an important role in the aging process of an entire organism, not just their individual cells. The implications of this are huge; not only may we better understand premature aging, but we may even be able to better control, and possibly reverse, the negative effects of aging in humans.

Telomerase has been studied for decades, and to understand what telomerase is and does, it's easiest to go back to the early 1970s. It was at this time that researchers found out that whenever a cell divides into two new cells, a tiny bit of the DNA at the ends of its chromosomes is lost. (Chromosomes are organized, string-like packages that store our DNA; people usually have 23 chromosome pairs in each cell.) Cells don't perfectly copy their DNA. Although it's a very small amount of DNA that's lost each division, it adds up over time and could eventually cause the loss of vital genetic information.

TELOMERES TO THE RESCUE

By the late 1970s, it was discovered how cells solve the problem of this lost DNA. When our cells divide, instead of losing valuable genes at the ends of our chromosomes, we just lose pieces of telomeres. A telomere isn't a gene (it doesn't have instructions for making anything in the cell), but is just a long piece of disposable DNA. (Specifically, telomeres are made up of a defined DNA sequence, TTAGGG, that is repeated several thousand times.) In this way, telomeres serve as a protective "cap" on the chromosome ends; when a cell divides, the telomeres are slowly shortened over time, keeping valuable genes from getting destroyed.

TELOMERASE

But even with these protective chromosome "caps," animals couldn't live for several generations with their telomeres continually shortening, especially if their shortened DNA was passed on to their offspring. About 50 to 200 base pairs of DNA are lost with each cell division, a process that clearly can't go on forever. Amazingly, somehow life finds a way to reverse this shortening: newborn babies have telomeres that are about 15-20 thousand DNA base pairs long, much longer than the telomeres of their parents.

This biological puzzle long perplexed researchers, until 1984 when Elizabeth Blackburn and Carol Greider (her Ph.D. student at the time) at the University of California, Berkeley, discovered an enzyme called telomerase. What does telomerase do? Telomerase is able to add telomere DNA back onto the ends of chromosomes, and consequently can preserve the integrity of the chromosomes over time.

But in humans only a few types of cells make telomerase: sperm and eggs ("germ" cells needed for reproduction), stem cells, and cells in regenerating tissues. These cell types must have a longer lifespan to pass genes onto a new generation, and to replace old or damaged cells over time.

The discoveries of telomeres and telomerase were such a breakthrough that Blackburn, Greider, and Jack Szostak (at Harvard Medical School) were awarded the Nobel Prize in Physiology or Medicine in 2009 for it. Blackburn also collaborated with Eduardo Orias, a professor at the University of California, Santa Barbara (UCSB), in the 1980s while investigating the genetics of *Tetrahymena*, a single-celled ciliated organism that Blackburn first discovered telomerase in. I had the great opportunity to work with Orias before starting graduate school at UCSB. Orias recently celebrated his 50th anniversary of working at UCSB, and continues to investigate *Tetrahymena* genetics.

A BIOLOGICAL CLOCK

In addition to protecting the ends of chromosomes, telomeres are also thought to function as a kind of clock for the cell, measuring how many times the cell has divided into two cells. When a cell has divided many, many times, and its telomeres have become very short, the cell detects this and stops itself from dividing ever again. It ceases to multiply and "senesces."

CANCER

What purpose would there be to stopping old cells from continuing to multiply? The answer is thought to be related to cancer; this telomere countdown may be an innate anti-cancer mechanism. To become cancerous, a cell needs to gain several mutations, and this process usually takes a long time to happen. Ideally, an old cell will die before its DNA has gained enough random mutations to make it cancerous, and it'll be replaced by a new cell.

But sometimes an old cell gets too many mutations before it can be eliminated (whether by chance or though exposure to carcinogens) and it becomes cancerous. Telomeres, telomerase, and cancer are intimately related, though, as with much in the field of cancer, we're still trying to figure a lot of these connections out. Cancer cells usually have short telomeres, probably because they were "old" cells that just barely escaped self-destruction. But cancer cells have figured out how to get around the problem of having short telomeres; unlike most of the normal tissues in the body, cancers make telomerase. This allows the cancer cells to maintain their telomere length and keep proliferating. They no longer have a self-destruct clock.

Consequently, there have been anti-cancer drugs developed that specifically block telomerase activity. Such drugs are in preclinical and clinical trials for treating prostate, lung, breast, pancreatic, and other cancers and have encouraging results. One such anti-telomerase drug, Imetelstat (GRN163L), developed by the biopharmaceuticals company Geron, showed promising results in preclinical trials against a wide variety of tumors, and has moved on to clinical trials testing the efficacy of this drug against non-small cell lung cancer, breast cancer, and other cancers.

TELOMERASE AND AGING

But while telomerase plays a deleterious role in creating cancers, it also plays a beneficial role in preventing, and possibly even reversing, aging. When telomeres are lost, tissues progressively atrophy, the body's supply of stem cells becomes depleted, healing is impaired, and organs can fail. Age-related diseases, such as atherosclerosis, Alzheimer's disease, and others, are associated with shortened telomeres. People born with shorter telomeres usually have a shortened lifespan. There are often other medical complications associated with shorter telomeres, but it's difficult to say, ultimately, whether the shortened telomeres are the cause of such conditions, or an effect. Also unclear is whether damage caused by shortened telomeres from aging can be reversed, or simply halted.

In the laboratory, it's been shown that human cells can have increased life spans if they're forced to make telomerase. These senescent-resistant cells also appear to be "normal;" they're not cancerous. This goes along with the thought that cancer cells' production of telomerase is an effect of these cells becoming cancerous, not the primary cause. However, while telomerase makes cells in a petri dish live longer, many scientists are skeptical as to whether increased telomerase production can make an entire animal live longer or be youthful.

But a recently published study in mice may have skeptics rethinking their stances.

THE FOUNTAIN OF YOUTH, AT LEAST FOR MICE

In November of 2010, a study published by Ronald DePinho and his postdoctoral fellow Mariela Jaskelioff at the Harvard Medical School in the journal *Nature* showed that degeneration in mice can be reversed by activating telomerase. To read the actual paper, see http://dx.doi.org/10.1038/nature09603.

The researchers created mice that lack telomerase, and consequently have short, dysfunctional telomeres. These adult mice had severe tissue degeneration, similar to premature aging, due to this mutation. To see whether telomerase could halt or reverse this degeneration, the researchers reactivated telomerase activity in these adult mice. The telomerase lengthened the short telomeres, caused cells to grow that had stopped growing, and eliminated the degeneration that had been previously seen in several organs, including some neurodegeneration. The telomerase didn't just stop more damage from occurring; it reversed the damage that had already been done.

THE FUTURE OF TELOMERASE

The exciting recent finding that telomerase may actually be able to reverse tissue damage in degenerative, adult mice has many implications. It may raise questions about using anti-telomerase drugs to fight cancer. At the same time, telomerase may be further investigated for possible use in reversing aging in people, although much work must be done before this is feasible; telomerase in mice and humans may function in different ways. But while further research will need to be conducted, this finding highlights the importance of telomerase, and how it may help us better understand and "treat" aging.

Further Reading

The following resources may be of interest for those wanting to learn more about telomerase and telomeres.

- Leonard Guarente, Linda Partridge, and Douglas Wallace's book *Molecular Biology of Aging*
- Catherine Brady's book *Elizabeth Blackburn and the Story of Telomeres*
- Mariela Jaskelioff et al.'s article "Telomerase reactivation reverses tissue degeneration in aged telomerase-deficient mice" at http://dx.doi.org/10.1038/nature09603

Viruses: Diseases and Cancer

Diseases, cancer, and medical uses related to viruses

Like many viruses, influenza (the flu) benignly lives and evolves in other animals. When people contact infected animals (such as ducks, which commonly harbor influenza), novel and potent viruses can cross over into the human population.

It has long been hard for people to accept that they can be hurt, or even killed, by something they cannot see. But this is exactly what viruses do; they evade our body's immune system, hijack our cells to create more viruses, and have consequently caused epidemics throughout human history. More recently, we've recognized they can also cause cancers, as well as other changes to our very genomes that we have yet to fully understand the implications of.

SMALLPOX

Possibly the oldest viral disease on record is smallpox, with accounts dating to 1100 B.C.E. in China and India. It made its way to Europe around 500 C.E., and then infamously followed the Europeans to the Americas in the early 1500s, destroying Native American populations. Because it is eradicated today, it can be hard to imagine just how difficult life with this disease was; of the people who caught it, up to forty percent died, mainly through damage to their heart, kidneys, or brain. People who survived could still have injured kidneys, disfiguring scars, or blindness (which it was the leading cause of).

Luckily, in the late 1700s, an observant British scientist named Edward Jenner noticed that dairymaids previously infected with "cowpox" (which caused red blisters on cow udders, and is similar to a mild case of smallpox in people) had immunity against smallpox. Using scabs from infected cow udders, Jenner infected several hundred thousand people with cowpox, effectively vaccinating them against smallpox. But even with a vaccine, like many viruses smallpox remained a significant health problem for years; in 1967, ten to fifteen million people were infected, and over two million died. The World Health Organization took action and created an intensive, $300 million program to finally put a stop to the disease, and it worked; it was officially eradicated in 1979.

MEASLES

Measles has caused epidemics in Europe and Asia for about two thousand years. Like smallpox, measles made its way to the Americas with early Europeans and wrecked havoc on the previously unexposed Native Americans. Given how contagious it is, its destructive ability is not so surprising; measles is possibly the most contagious virus in the world, even more than the common cold, as a person can catch it just by breathing air in a room where an infected person was two hours earlier. And it can kill about ten percent of the people it infects, if they face malnutrition and inadequate health care, with weeks of high fevers, rashes, severe diarrhea, dehydration, and other infections. Trying to survive measles was simply a part of life for much of human history. In 1963, a vaccine was developed and the disease mostly vanished in developed countries. Like smallpox, measles too could be completely eradicated from the human population if a global, stringent vaccination policy were carried out for several years because, unlike many diseases, measles doesn't live in any other animals we know of; we're its only home. While the number of lives measles claims each year is declining, killing about 873,000 people in 1999 down to 164,000 in 2008, it remains a significant cause of mortality for children in many developing countries.

HERPES

Like measles, herpes (caused by the herpes simplex virus 1 and 2) has also caused known epidemics for about two thousand years. When the virus multiplies, it creates blistering lesions, which spread across the infected area and gave the virus its name; herpes is Greek for "to creep." While simplex 1 targets the skin and mucous on the lips and mouth, creating "cold sores," simplex 2 attacks these kinds of tissues in the genitals. It's been long known that the virus spreads through the contact of these sores; the Roman emperor Tiberius Julius Caesar Augustus banned kissing possibly in efforts to prevent the rampant spread of herpes throughout Rome.

While the body's immune system eventually fights off herpes, it isn't gone for good. A persistent viral infection, herpes simply retreats when "beaten," waiting to return and fight another day. Where does herpes retreat to? It actually hides in the body's nerve cells, sometimes for years before returning to cause lesions. While there's no cure yet, there are treatments to control its severity and several vaccines in trials.

The virus that causes chickenpox (the varicella-zoster virus) belongs to the same family as herpes simplex, and retreats in a similar manner. After causing chickenpox in children, the virus retreats to nerve cells (in the vertebrae), but can reemerge years or decades later and cause the painful rash condition of shingles.

The Epstein-Barr virus, which causes mononucleosis ("mono") and may infect over ninety five percent of all people at some point, also belongs to the herpes family. Like chickenpox, it can make a repeat performance later in life, particularly in the form of cancer; it's been associated with causing Burkitt's lymphoma, nasopharyngeal carcinoma, and several other cancers.

INFLUENZA

While it is hard to track the historic spread of influenza (also known as the flu) as its symptoms resemble a respiratory disease, it's thought that, along with smallpox and measles, it too made its way from Europe to the Americas in the early 1500s. Altogether, the three viruses killed as much as ninety-five percent of some Native American tribes.

But the worst influenza epidemic was yet to come; in 1918, half of the world came down with influenza. Within months, between forty and one hundred million people (about five percent of the world's population), mostly healthy young adults, died from the "Spanish flu." It was the worst pandemic in modern history. It's also why scientists and physicians were terrified in 2009; the influenza strain of 1918 was also H1N1. Luckily, the 2009 H1N1 influenza strain, the "swine flu," did not prove to be as virulent as the 1918 strain.

So why is the flu so effective, and why is there a different flu vaccine every year? The protein "coat" that surrounds the influenza virus is what our immune system uses to recognize and attack the virus, but the virus changes this coat all the time, making it hard for our immune system to identify. Part of why it changes all the time is because it doesn't just live in humans; influenza often benignly lives in the digestive tract of pigs, ducks, and other waterfowl. There it swaps genes with other influenza strains, and occasionally a very potent virus is created this way. Consequently, where people are often in contact with these animals new influenza strains often find their way into the human population, such as in many Asian countries.

RABIES

While people in the U.S. today may be remotely familiar with rabies from the 1957 Walt Disney Classic "Old Yeller," the disease was a serious threat to everyday life until the early 1900s. After being bitten by an infected animal, a person often developed rabies, which included a fever followed by depression, restlessness, frothing, unquenchable thirst, and certain death, if untreated. Luckily, the brilliant French chemist Louis Pasteur developed a vaccine (using dried fluids from infected animals' spinal cords, where the virus resides). Unlike most vaccines, the rabies vaccine can be used after a person has been in contact with an infectious source because the virus can take a week to over a year to infect a person after exposure. In 1885, Pasteur first successfully used his rabies vaccine on a 9-year-old boy who was attacked by a rabid dog. Within the next couple years, Pasteur's vaccine saved about 2,500 lives, and has saved many more since. While rabies is rare in the U.S., it is still common in developing countries, killing around forty to seventy thousand people annually.

HEPATITIS

Around the same time that Pasteur was curing rabies, people were inadvertently spreading hepatitis while reusing immunization needles to vaccinate against other viruses. Hepatitis had yet to be identified, and people did not know that they were developing yellowish skin (jaundice) due to viral liver damage. This symptom is actually caused by three different viruses, hepatitis A, B, and C. Hepatitis A (HPA) is the mildest, spreading orally from bad food or water in unclean living environments and causing nausea, vomiting, and jaundice for a month or two, but then allowing for complete recovery.

Hepatitis B (HPB) is much more serious. HPB is very infectious, spreads through sexual contact or body fluids, and is often deadly. Worldwide, about three hundred million people (five percent) have a chronic infection, mostly in Southeast Asia and tropical Africa, but this also includes 1.25 million Americans. Around one million people die from HPB annually, due to liver disease (cirrhosis) and the five million liver cancer cases it causes. Luckily, a vaccine was developed in the 1970s, making it the first cancer vaccine. (This cancer vaccine was discussed earlier in this chapter, in "Cancer Vaccines, Part I.")

It took researchers a bit longer to identify hepatitis C (HPC), the most deadly of the hepatitis viruses in the U.S. Isolated in 1989, HPC was found to also cause liver disease and cancer. In the U.S., it infects nearly four million people, occurring more frequently than any other blood-borne infection here, and killed ten thousand in 2001. While many drugs can be used to treat it, and it's become the leading cause of liver transplants in the U.S., unfortunately, because HPC mutates so quickly, a successful vaccine has yet to be developed.

POLIOVIRUS

While HPC remains a significant problem for many in the U.S., polio (caused by the poliovirus) has mostly been eradicated from this country. 1916 witnessed the first U.S. polio epidemic, and it remained a significant fear for four decades. Like hepatitis A, polio spreads through contaminated, swallowed substances, such as from swimming in a public swimming pool. While most infections go unnoticed, as the virus stays in the intestines, sometimes it enters the nervous system and proceeds to kill neurons, often resulting in paralysis. After realizing this in the late 1940s, American medical researchers Dr. Jonas Salk and Dr. Albert Sabin developed different vaccines. While it is now rare in the U.S., polio is still a major concern in South Asia and sub-Saharan Africa, where, in 2000, nearly four thousand were afflicted with pain and paralysis, and hundreds of thousands more were infected. Efforts to perform a global eradication are underway, but it has been difficult to implement vaccination programs in countries with social and political upheaval.

EBOLA

Ebola has caused many of the most recent viral epidemics. Discovered near the Ebola river valley in Zaire in 1976, Ebola killed over ninety percent of people in over fifty villages there in a fast, gruesome manner, with hemorrhagic fevers. The virus is transmitted to people from animals (most likely infected carcasses), and then spread between people through the contact of contaminated bodily fluids. Although there have been multiple epidemics since 1976, because the virus is actually fragile without a host, and kills a person within a week, the epidemics have remained fairly localized. While there is no vaccine available or standard treatment, several vaccines are in trials.

HIV

Human immunodeficiency virus (HIV-1), the virus that causes acquired immunodeficiency syndrome (AIDS), is currently a pandemic that has infected nearly forty million people, and has killed over twenty five million people since it was recognized in

1981. HIV-1 attacks the host's immune system, causing a person to die from opportunistic infections. Although HIV-1 is from a chimpanzee virus (the simian immunodeficiency virus [SIV]), SIV does not cause a disease in the chimpanzee, but it does in other non-human primates. While anti-viral treatments are available, vaccines are not, although there have been promising stem cell findings based on individuals who are somewhat resistant to HIV-1 infection (see Chapter 1 for details). HIV-1 is an ongoing, significant problem, with nearly three million new cases reported in 2007 and two million deaths. For now, perhaps the best weapon against HIV-1 infection is awareness.

CANCER

Although we've known that viruses can cause cancer since 1911 (with the discovery of the Rous sarcoma virus, a chicken virus), the idea has only recently been widely accepted by the scientific community. Because many viruses reproduce by inserting their genes into our genomes, they can cause cancer; these insertions can (usually randomly) disrupt important genes we have for preventing cancer, or insert their own cancer-promoting genes. In addition to the Epstein Barr virus and hepatitis mentioned above, another prominent example is the human papillomavirus (HPV), which causes cervical cancer and kills a quarter million people globally (four thousand in the U.S.). Luckily, a vaccine against HPV, and consequently against cervical cancer, has been developed. (This cancer vaccine was discussed earlier in this chapter, in "Cancer Vaccines, Part I.")

Because viruses have a habit of inserting their genes into our genomes, it may not be surprising to learn that our genome is actually eight percent viral DNA, specifically belonging to bornaviruses. For a discussion on this fascinating topic, see http://www.wired.com/wiredscience/2010/01/bornavirus-in-human-dna/. These RNA viruses first started their invasion over forty million years ago. Today, it's unclear what the implications are for our genome; some of the viral genes surprisingly appear to be beneficial, while others may not be (and may even cause schizophrenia).

Either way, scientists have been using viruses in research for decades, such as in studying how different, introduced genes behave in other organisms, or, more recently, in promising new gene therapies to treat genetic diseases. For a discussion on these gene therapies, see http://singularityhub.com/2010/09/23/man-with-blood-disease-free-from-life-of-transfusions-thanks-to-gene-therapy/. It appears as though the tables may be starting to turn; instead of viruses using us to their advantage, as they've done for millions of years, we're learning how to apply their unique abilities to further our knowledge and ourselves.

Further Reading

The following resources may be of interest for those wanting to learn more about viruses and cancer:

- Luis P. Villarreal's book *Viruses and the Evolution of Life*
- Bruce A. Voyles and Barry E. Zimmerman's book *The Biology of Viruses*
- David J. Zimmerman's book *Microbes and Diseases That Threaten Humanity*

CHAPTER 5

WHAT YOU EAT

Alcoholic Origins

From food to fermentation

Grape vines have been cultivated since 6000 B.C.E., and while many are familiar with the production of wine from grapes, people have created other alcoholic beverages from an astonishing array of food products locally available to them.

Since the beginnings of civilization, people have created alcoholic beverages from the fermentation of food found nearby. While some alcoholic beverages have familiar origins, such as grapes for wine and grains for beer, others are more unusual, but all reflect how people constantly make use of their environment.

ITS ORIGINS: GOOD FOR WHAT AILS YOU

How did alcohol become the widely accepted part of life that it is today? Drinking alcoholic beverages has distinct advantages over drinking water, especially for early man. Alcohol kills pathogenic microbes, which is quite important when other available water sources might be contaminated. In further support of this justification, it's been found that some ancient beer had significant amounts of the antibiotic tetracycline, a naturally occurring antibiotic made by some bacteria. This may have caused beer consumption to result in good health. Alcohol is also a psychoactive drug and depressant, which may have helped it gain the central role in social interactions that it maintains to this day.

HOW IT CAME ABOUT

The oldest alcoholic beverages, dating as far back as 6000 B.C.E., were made from fermented grains, grapes, honey, or milk. Fermentation is a naturally occurring process by which yeast converts sugar into ethanol (ethyl alcohol); food with a lot of sugar ferments into alcohol readily, often by simply exposing the food to air. Different alcoholic beverages arose in different places around the world based largely on the foods commonly on hand to people living there.

Despite its widespread popularity today, beer was most likely not the first alcoholic beverage developed; a very involved process is required to produce beer. More likely (although the topic is hotly debated) wine, mead, and kvass were the first alcoholic drinks created by people. They are made from grapes, honey, and milk, respectively, using a simpler fermentation process.

WINE

In the regions of Southern Europe, the Mediterranean, the Middle East, Asia, and North America, wild grapes grow well and consequently were an important source for creating alcohol (although some theorize that date palms were actually the first fruits used to make wine). For climatic reasons, the best areas for growing wine grapes are between 30 and 50 degrees North and South latitude. This is indeed where the first wines were developed. In the Middle East, people were cultivating grape vines by 4000 B.C.E., although in Georgia this went on as far back as 6000 B.C.E. Using the fruit to make alcoholic beverages was a natural next step.

In brief, to create wine, the ripe grapes are quickly picked, crushed, fermented, and then aged. For white wines, the skins and seeds are removed when crushed, while for red wines the skins and seeds are left in. Most early wines were reds. Because the yeast *Saccharomyces cerevisiae* naturally lives on grapes, there is no need to add yeast to start the fermentation process. However, most wine makers today add yeast (often of a different species) to achieve predictability in their wine. At the end of the process, one metric ton of grapes is transformed into 140 to 160 gallons of wine.

Viniculture (cultivating grapes to make wine) technology was passed on from civilization to civilization through trade and conquest. Around 3000 to 2500 B.C.E., it found its way to the Egyptians, but because wine was four times as expensive as beer it was not as frequently consumed. From the Egyptians it spread to the Greeks, who developed locally-made wines around 1700 B.C.E. Like most of Greek culture, viniculture found its way to the Romans, where the technology was catalogued. Nearly 2,000 years ago, Pliny the Elder classified grapes based on their ability to resist disease, their soil preferences, time to ripen, wines produced, and more. From the Romans, viniculture spread to cultures throughout southern Europe (modern-day Italy, Spain, France, and Germany).

When the Roman Empire collapsed, monasteries maintained the brewing and wine-making techniques as an essential part of religious ceremonies, harboring viniculture for nearly 1,300 years. Medieval monks in France developed a variety of wines, from Burgundies and Bordeaux to Champagnes, and more. Indeed, it was monks, of the Franciscan order, who established the first Californian missions in the 1700s, and with them the California wine industry.

BEER

While wine is generally considered easier to make than beer, beer has one big advantage; Beer's main ingredient, cereal grains, are not so picky about where they grow. Consequently, beer is often made in places where wine cannot be, which is one reason why beer is cheaper. Additionally, grain has been a primary component of peoples' diets for thousands of years; barley and wheat were cultivated around 6000 B.C.E. in western and central Eurasia, East Asia, and Central America. Beer brewing most likely first took place in Mesopotamia (what is today Iraq), and, much like wine, spread throughout the trading cultures.

Beer can be made from a variety of cereal grains, including oats, rye, corn, millet, rice, and others, although barley has always been most commonly used. Nearly 150 million tons of barley are harvested every year, in over 100 countries. These grains contain lots of starch that would normally feed the growing plant. The brewer's job is to turn this starch into alcohol. Because yeast turn sugar into alcohol, the starch must first be converted into sugar that the yeast can eat (unlike with making wine, where grapes are basically bags of sugar).

To first help convert the barley's starch into sugar that can be fermented, the barley undergoes a process called "malting." This involves soaking the barley in water, which stimulates the seeds to germinate and start growing. They become softer and even sprout rootlets. More importantly, during germination the seeds make enzymes that will be needed to break the starch down to sugar. However, germination is abruptly stopped by heating and drying the seeds, which ensures that starch remains and nutrients are not wasted. How exactly the malting process is done affects the beer's ultimate color and flavor; the more the malt is heated, the darker its color. This is why ales are usually darker than lagers.

After malting, water as hot as 149 degrees Fahrenheit is added (in a process called "mashing") to melt the starch. At this point, about 80% of the liquid, now called "wort," is ready to be fermented. While the leftover, solid grain material (now mostly hulls) is turned into cattle feed, the wort is boiled with hops (the flower of the hop plant). In many European countries, the addition of hops did not become popular until the early 1300s.

Yeast is then finally added to ferment the cooled wort. Ales and lagers use different yeast species (*S. cerevisiae* and *S. pastorianus*, respectively). The selection of yeast strain is more important with blander beers, because the yeast flavor will not be tasted much over the flavor of a bitter beer. And because cereal grains do not normally harbor yeast, early on, people found that the fermentation of grains could be stimulated by adding fruit or by putting the boiled grains in a container that previously held fermented grains.

ANCIENT ALCOHOLIC DRINKS NOT FROM PLANTS: MEAD AND KVASS

Before cereal grains were even cultivated, it's thought that in northern and western Europe honey was used to create the first alcoholic beverages; namely mead, also called honey wine. There's evidence that honey-based alcoholic beverages were also anciently produced in Africa and Asia. It was the first popular Greek drink. While honey is basically fermented to create honey wine, many variations have developed, which include adding spices, hops, or herbs. Kvass, an alcoholic beverage made from fermented rye bread or milk, is also thought to be one of the oldest alcoholic drinks, possibly predating beer.

CIDER

Mild alcoholic cider is created by fermenting apple juice, and consequently it has been historically popular where apples are abundant. During the Early Middle Ages (500 to 1000 C.E.), cider became a popular local drink with the Germans and Celts. In the United States, cider was by far the most popular drink until the early 1800s. Fruitful apple orchards across New England and New York turned out "apple wine" that was often drunk with every meal, and even between them.

THE EUROPEAN ALCOHOLIC EXPLOSION OF THE 1600S: BRANDY, WHISKEY, GIN, AND RUM

In the 1600s, a variety of alcoholic beverages emerged in Europe and were embraced to varying degrees and for different usages. Scottish and Irish whiskey, created from fermented grain mash, made its way around Europe, but its harsh taste prevented it from being widely accepted. Instead, the drink of choice for the rest of Europe became gin, which was introduced by the Dutch in 1672. The Dutch created gin by taking Scottish grain alcohol and adding juniper berries. Brandy, made by distilling wine, also surfaced at this time, although it was only used medicinally.

Shortly after 1650, a new alcoholic drink was refined that would dominate trade routes for over a century: rum. To make rum, molasses, a by-product of sugar cane refinement, is fermented into alcohol, and then distilled (a relatively recent invention) and concentrated. West Indian molasses was imported to New England where distillation turned it into rum. The rum was then exchanged for African slaves, who were sold for molasses, completing the "Triangle Trade."

MORE LATIN AMERICAN DRINKS: FROM THE AGAVE CACTUS TO COMMUNAL CORN CHEWING

Aside from rum, other alcoholic beverages produced in Latin America include mescal, pulque (a drink of the Aztecs), and tequila, which are from the juice of the agave cactus. Mescal can also be distilled from the peyote cactus, which can cause mild hallucinogenic effects. Chichi is often made from fermenting apples, strawberries, yucca, and oats – and, more interestingly, can also be made from chewing corn in groups. Chewing the corn introduces salivary enzymes that can help create sugars for fermentation, and so several people chew the corn and then deposit it in a communal bucket to ferment.

SAKE

Though agave cacti have never been very common in Japan, rice is omnipresent. Sake, a wine made from rice, is prepared more like a beer than a wine, as the starchy seeds must first be converted to sugar in order for fermentation to take place.

While this history barely touches on the multitude of alcoholic beverages that humans have developed, it highlights not only how inventive people have been at creating drinkable products from their surroundings for millennia, but also how amazingly flexible the fermentation process is at turning plants or animal by-products into alcohol.

Further Reading

The following resources may be of interest for those wanting to learn more about the history and biology of alcoholic beverages:

- Charles Bamforth's book *Grape vs. Grain: A Historical, Technological, and Social Comparison of Wine and Beer*
- Thomas Babor's book *The Encyclopedia of Psychoactive Drugs: Alcohol, Customs and Rituals*

Invigorating Leaves: The History and Biology of Tea

How people have cultivated and studied a humble plant for millennia

After millennia of careful cultivation and experimentation, today hundreds of varieties of tea are made from the tea "tree," *Camellia sinensis*.

When thinking of tea, it's easy to picture dozens of varieties, packed in colorful boxes, readily available on a local grocery store's shelves. But this is just a small glimpse of the current tea industry. Tea is second only to water in terms of global consumption, and there is an astonishing 350 to 500 varieties, all thanks to a single plant species: *Camellia sinensis*. This enormous variety of teas is due to millennia of cultivating *C. sinensis*, studying how different climates and soils affect its growth, and investigating different ways to prepare its leaves for consumption.

CAMELLIA SINENSIS

The plant that tea comes from, *Camellia sinensis*, is a small evergreen that probably originated around southwestern China (specifically on the Yunnan and Guizhou plateaus, though this is hotly contested). It's been living there for a very long time, dating back to the Tertiary period (65 to 2.6 million years ago), and luckily survived the great ice ages because the glaciers didn't reach these plateaus. *C. sinensis* continued to spread throughout southeastern Asia, thriving in certain mountainous areas.

The plants have small, rose-like white flowers and can grow into towering trees, often up to 40 feet tall, with records of 108-foot-tall living behemoths that may be over 1,700 years old. However, to keep them a feasible height for picking leaves for tea, *C. sinensis* is carefully pruned down to a two- or three-foot-tall bush. Two main varieties are generally recognized, the "China bush" (*Camellia sinensis* var. *sinensis*) and the "Assam bush" (var. *assamica*). These varieties look rather different from each other physically (the China bush grows more like a bush and the Assam bush more like a tree) and prefer different conditions for ideal growth (the China bush prospers on cool mountainsides and the Assam bush prefers jungle-like areas), but the differences in their genetics are mostly negligible.

THE CREATION OF TEA

It will probably never be known for certain who first made a drink from leaves of the tea plant, but legends exist and some educated guesses for how it most likely came about have been made. Some hypothesize that prehistoric people may have found wild tea trees and, looking for food to eat, could have chewed on the leaves and found that this gave them more energy to carry out their daily routines. Perhaps they even started to tend to some of these plants. What are thought to be ancient tea "gardens" have been discovered (specifically growing in the depths of the Brahmaputra valley), but it's hard to know how old such gardens are.

Some Chinese legends credit tea's accidental discovery to Shennong (known as "the Divine Cultivator" because he's also credited with inventing agriculture). As the legend goes, around 3000 B.C.E. Shennong was boiling water outside and leaves from the tea plant fell into his water, creating a drink he enjoyed and thought to have medicinal properties.

During the Shang dynasty (1766 to 1050 B.C.E.), fresh tea leaves were boiled in water with other plant materials, including leaves, barks, and seeds, to create medicinal brews that were used to treat a variety of ailments. By about 300 B.C.E., tea was prepared for the first time without other plant material and consumed not necessarily as a medicinal tonic, but as an invigorating, bitter drink. Drinking tea quickly became popular among monks and priests of Buddhism, Taoism, and Confucianism, as they found it helped them stay awake while meditating for long periods of time. They were enthusiastic supporters, proclaiming that all people should drink tea. As more and more people drank it on a daily basis, leaves were often dried and processed to make tea, as it was becoming unfeasible to meet increased demands using fresh leaves.

In these early days of tea drinking, the event that helped spread its popularity throughout China more than anything else may have been the building of the Great Wall. As workers were brought from across China to create the Great Wall under Emperor Qin Shihuangdi (221 to 210 B.C.E.), information quickly spread about relatively localized customs, such as tea drinking, which was previously popular mostly in the western provinces. Soon, tea drinking was trendy throughout the country. Around 50 B.C.E. tea plants began being cultivated for the first time to supply the desired leaves.

AN EVOLVING DRINK

Over the next several centuries, tea went through many metamorphoses as prevailing tastes changed and challenges arose. The first major transformation, which took place between 200 and 600 C.E., changed tea from a bitter and rather crude drink to a sweeter beverage – though not sweet by today's standards! Until this point, tea leaves were dried and then charred before boiling them to make tea, but people found that if they instead steamed the leaves, the pliant product could be compressed into little tea cakes which they then baked. The baked cakes kept well, and small bits of tea could be chipped off at a time and boiled in water for tea. These tea "bricks" became a common form of currency, and were so essential to people living outside of China that they traded horses for them.

But more importantly, these cakes made tea that was much less bitter than tea produced by the dried, charred leaves. This made tea-drinking a pleasurable activity, which was elaborately refined during the Tang dynasty (618 to 907 C.E.). Also during this time, tea and the current tea-drinking customs in China made their way to Japan for the first time, where they took on a life of their own. However, tea was not enjoyed beyond the Japanese emperor and his court until centuries later, around 1200 C.E., when Zen priests planted many tea plants in Japan.

During the Song dynasty (960 to 1279 C.E.), steamed tea leaves were more finely ground to create cakes made of powdered tea. This further refinement changed how tea was commonly served; now the powder could be scraped straight into the drinking bowl and whipped with boiling water to create a beautiful green froth. At this time teahouses became popular, providing places to not only share tea and snacks in public, but to socialize, gossip, play games, carry out business, and more.

But still tea kept on changing; by the end of the Song dynasty, people were experimenting with making tea from loose leaves, which made it easier to measure out. When Kublai Khan and the Mongol Empire took over China (1271 to 1368 C.E.), they further explored the new loose-leaf technology. However, many still preferred their tea powdered because of its supposed superior flavor.

After the Mongols, the Ming dynasty (1368 to 1644 C.E.) developed two new major types of tea: black tea and scented teas. The tea brewed until this time was primarily what we'd know today as green tea, but during the Ming dynasty it was found that black tea could be made by oxidizing the leaves. (Technically, green tea is also oxidized to some extant, but not nearly as much as black tea is.) Although the Chinese preferred green tea over black tea, black tea made better tea bricks; they could withstand travel over long distances without getting ruined from extreme temperatures or mold. This made black tea ideal for exportation. Using flower petals, such as from jasmine and rose, to make perfumed teas also became popular at this time.

TEA GOES TO EUROPE

During this experimental time, tea made its way from China to Europe through Dutch traders around 1610 C.E. It's thought that Europeans added milk to their tea because they wanted to imitate how they heard the emperor of China drank his tea. However, the emperor of the time was a Manchu tribesmen of the Qing dynasty (1644 to 1911 C.E.), and the tribesmen preferred black tea with fermented mare's milk added to it. (This is also how the Mongols drank their tea.) Of course, the rest of China still preferred green tea without milk, but perhaps the Europeans were only interested in the Emperor's habits. In addition to cream, the British also added sugar, another new import, making their twice-daily tea drinking routine quite the reenergizing experience.

U.S. INNOVATIONS

Two major tea "improvements" took place in the 1900s C.E. in the U.S: the tea bag and iced tea. In 1904, the World's Fair in St. Louis was a real scorcher, and tea vendors with hot Indian black tea did not fare well. However, tea plantation owner Richard Blechynden had a great idea: Pour the tea over ice. Thus, iced tea was born, and has become increasingly popular in the U.S., as today over 80% of tea consumed in the U.S. is iced (mostly black) tea.

The invention of the tea bag came about in a similarly unplanned manner. In 1908, Thomas Sullivan, a tea importer in New York, sent out tea samples in little silk bags. When people found they could make tea with the leaves still in the bag, demand rose for it, though many tea connoisseurs still think that tea bags make lower quality tea.

TEA TODAY

Tea is now produced in more than 45 countries, spanning several continents; it's hard to meet the demand for a beverage that's second only to water! There are now hundreds of different teas available for the interested consumer, and a variety of groups of tea, although the four main groups are green tea, black tea, white tea, and oolong tea.

These groups of tea differ mainly by where the plants are grown and how the leaves have been prepared. The tea leaves are picked (usually the top two leaves and a bud at a time) primarily during the spring, but they can be picked during the rest of the season depending on where they are grown and the kind of tea that will be made. To make green tea, the leaves are steamed (or heated in some other way, such as on hot woks) and then rolled or dried. (This allows the leaves to be only minimally oxidized, with the heat halting the oxidation process.) To make black tea the leaves are sometimes rolled or cut, and then allowed to undergo full oxidation (the leaves are left to turn brown) before drying. Some consider black tea to be an extension upon making oolong tea, which also requires some bruising of the leaves, but then only a short oxidation step before drying. Likewise, green tea is similar to white tea in that they both skip the extensive oxidation step, but leaves for white tea are not rolled. The full oxidation step is what makes black tea have more color, but less astringency, than green tea. Often many different varieties are blended together to create the final store-bought product. For example, most black teas actually contain a few dozen different black teas made in a couple different countries.

HERBAL TEAS

Herbal "teas" have also become very popular, though technically they're not a "tea" but are an herbal "infusion" or "tisane" because they do not contain leaves of the *Camellia sinensis* plant. Herbal "teas" instead contain different herbs, leaves, dried fruits, flowers, and more. Although strictly herbal "teas" do not contain *C. sinensis*, boiling various plant matter with tea leaves to create rich brews is a long tradition. Around 200 to 600 C.E., such concoctions frequently included onions, orange, and ginger, and later during the Tang dynasty (618 to 907 C.E.) cloves and peppermint, and the occasional delicate flower or fruit for court ladies, were added to this mixture. Still later, the Song dynasty (960 to 1279 C.E.) experimented with adding different juices, including pickle juice and, a particularly sweet treat, plum juice. While many parts of Europe today enjoy cream in their tea, this was not how it was originally prepared in Europe, but spices such as nutmeg, ginger, saffron, and salt were added instead. For the avid tea drinker, perhaps these historical examples will give inspiration for the next experimental brew.

TEA AND YOUR HEALTH

In the next section of this two-part series we'll discuss what exactly is the biochemistry that takes place in a good cup of tea, how this varies from one type of tea to another, and the hotly debated controversy of what drinking tea can (and can't) do for one's health.

Further Reading

The following resources may be of interest for those wanting to learn more about tea and its historical biology:

- Mary Lou Heiss and Robert J. Heiss' book *The Story of Tea*
- Kit Chow and Ione Kramer's book *All the Tea in China*
- Bennett Alan Weinberg and Bonnie K. Bealer's book *The World of Caffeine*
- Yukihiko Hara's book *Green Tea: Health Benefits and Applications*

Special thanks to Marlon Molinare of Chan Teas for feedback on this essay.

To Your Health: The Biochemistry of Tea

How the biochemistry of tea gives it its unique color, aroma, and flavor, and possible healthful attributes

A series of complex chemical reactions turn leaves from the tea plant (*Camellia sinensis*) into a delicious cup of tea.

There are hundreds of varieties of tea today, all created from a single plant species: *Camellia sinensis*. People have spent millennia cultivating the tea plant and discovering how to alter the chemistry of its leaves to create different types of teas. In the previous section of this two-part series we explored the history of tea. In this section we'll venture into the realm of biochemistry to reveal what gives different teas their distinctive flavor, appearance, and potential health benefits.

IT'S IN THE LEAVES

Camellia sinensis, like any plant or living organism, is a complex mixture of many different chemicals. While there is still much we don't understand about all of the chemicals and chemistry that goes on to make a delicious cup of tea, scientists have discovered many chemicals present in tea leaves that contribute noticeably to different aspects of tea.

Much of it comes down to how the leaves are treated. To look at this, it's easiest to focus on the two most commonly consumed types of tea: black tea and green tea. To make green tea, fresh tea leaves are steamed (or heated in some other way, such as on hot woks), then rolled or dried. As was mentioned in the previous section, this lets the leaves be only minimally oxidized, with the heat halting the oxidation process. For black tea, partially withered leaves are rolled or cut, then allowed to completely undergo oxidation (the leaves are left to turn brown), and then are dried.

These processing differences are what make green tea and black tea distinct from each other. Specifically, the chemicals found in green tea are very similar to the ones found in the leaves of the tea plant naturally, whereas because the leaves undergo complete oxidation to create black tea, black tea contains a variety of chemicals not found in the fresh leaves or green tea. One of the major chemical groups this oxidation affects is the catechins.

GREEN TEA AND CATECHINS

Polyphenols are chemicals that are abundant in tea leaves and are known to cause many of the properties we enjoy in tea. In nature, members of this large group of chemicals are found mostly in plants and act as antioxidants. (More on the topic of antioxidants to come.) Most of the polyphenols in tea leaves (about three fourths of them) belong to a more specific subgroup called catechins. Several different catechins are present in the leaves, and in the green tea itself they account for up to 30% of the weight of dry tea. It's this high level of catechins that gives green tea its characteristic bitter and astringent taste.

BLACK TEA

Black tea, however, has a much lower amount of catechins (they only make up about 10 to 12% of its mass) because of the oxidation process; when the leaves are rolled or crushed and left to oxidize to make black tea, chemicals in the leaves that were previously separated from each other are released and can now mix and trigger reactions. Because of this, during the complete oxidation process most catechins in the leaves undergo a chemical reaction to become other chemicals (thanks to an enzyme called polyphenol oxidase). Specifically, the catechins couple together to form chemicals called theaflavins and thearubigins. (They're polyphenols too.) Most of the theaflavins formed actually turn into thearubigins; ultimately, dried black tea has about 1 to 6% theaflavins and 10 to 20% thearubigins. These chemicals are not present in green tea. In fact, the enzymes needed to make these chemicals mostly become inactivated in the green tea by heat when the leaves are steamed (halting the oxidation process early on).

So what do these "thea-whats-its" do to black tea? Theaflavins are what make black tea a darker, reddish-orange color; they're molecules of pigment. Thearubigins also contribute to black tea's color, as they are orange-brown pigments. In addition to color, theaflavins also add to flavor; they are part of what causes black tea to be less astringent than green tea (another key factor is the decrease of catechins in black tea).

But there's a lot of other complex chemistry that goes into making black tea; just the tip of the iceberg is the many cascading chemical reactions that catechins are thought to undergo, where one chemical product reacts with others, creating the rich diversity of chemicals found in black tea. The chemistry of a lot of these, such as thearubigins, is still not well understood.

AN APPETIZING GREEN

Green tea, lacking pigment from theaflavins and thearubigins, has a green color thought to be due to several different chemicals in the fresh leaves, including chlorophyll, carotenoids, and several different flavonols. Chlorophyll gives green tea a bright green coloring, and is just another reason to make sure to drink that tea before it's been stored too long; chlorophyll, like other essential chemicals in tea, degrades over time, especially at high temperatures and in humid conditions. This is why old tea won't be as vividly green as fresh tea. Flavonols, though making up less than 1% of the mass of the dry tea, nevertheless contribute a bright yellowish green hue and even add to the astringency of green tea by reacting with caffeine.

CAFFEINE

Speaking of caffeine; it's probably the most widely-known chemical in tea, and acts as a stimulant of the central nervous system and a contributor to tea's bitter taste. Although dried tea has more caffeine than the same amount of dried coffee, because much less tea is used when making a drink, many may be surprised to learn that a cup of coffee actually has a lot more caffeine than a cup of tea. While a five ounce cup of drip coffee has 60 to 180 milligrams (mg) of caffeine, a same-size cup of black tea only has 25 to 110 mg caffeine, and green tea has even less, at 8 to 36 mg. That said, the exact amount of caffeine in any given cup of tea is very variable, and can be affected by a number of factors, such as the specific environment the plant was grown in, which leaves were used from the plant, and how hot and long the leaves are steeped for. And caffeine is yet another reason to drink that tea fresh; if black tea is stored too long, the caffeine is thought to degrade and make the tea bitterer.

THEANINE AND RELAXATION

But the stimulating effects of even the relatively low levels of caffeine in tea are curbed by the presence of a chemical called theanine. In nature, theanine (also known as y-L-glutamyl ethlamide) is very rare. In the early 1950s it was first found in tea leaves, but outside of tea leaves, theanine has only been naturally found in one species of mushroom (*Xerocomus basius*). When tea is consumed, the theanine increases dopamine levels and alters levels of serotonin in the brain, and overall causes a person to feel very relaxed, yet mentally alert, within about 30 to 40 minutes of drinking it. Theanine also greatly adds to the sweet and savory taste of tea, countering and building upon the bitterness of the caffeine and other chemicals. And this induced sense of relaxation is highly prized; usually a high-quality tea has a greater amount of theanine, up to about 2%.

WEIGHT LOSS?

Theanine combined with caffeine, and catechins alone, are also suspected to work to decrease body weight. Catechins are thought to interact with enzymes involved in "burning" fat to do this. In experiments where people were given catechin extracts for three months, it was found that people taking the extracts lost slightly more weight than those who weren't. But these catechin doses were at levels far higher than those found in normal tea; it's still unclear whether drinking tea normally and casually significantly affects weight.

ANTIOXIDANTS

Many of the chemicals in tea discussed above have antioxidant properties: catechins, theaflavins, and thearubigins in particular. Catechins (particularly the main catechin in tea, epigallocatechin gallate, or EGCG) are thought to be even better antioxidants than vitamins C and E. But this isn't so surprising; lots of antioxidants are in fruits and vegetables in general. Antioxidants have received a lot of attention in recent years because they may reduce the risk of different diseases, such as cardiovascular disease and cancer, and are even thought to potentially decrease aging.

But how well-founded are these claims? In the 1950s it was found that our bodies withstand damage all the time inflicted upon us by natural reactive molecules called free radicals (also known as reactive oxygen species). These energetic free radicals can run into other molecules in our cells and damage them. Researchers thought that our bodies must take countermeasures to fight these free radicals through "antioxidants." But in the last few decades, researchers have been testing whether antioxidants can combat the diseases mentioned above and only a few studies have shown any encouraging results, while the rest (the majority) didn't show antioxidant supplements to be at all beneficial. While the antioxidant theory may not be completely off, it is most likely much more complicated than we think it is; many factors are involved in any given diet. But eating lots of fruits and vegetables, even if antioxidants aren't a miracle cure, is still probably a very safe bet for a healthy lifestyle.

DRINKING TO YOUR HEALTH

Drinking tea is thought to help reduce the risk of a number of other diseases and health conditions in addition to the ones discussed above. This includes improving the function of the central nervous system, kidney, liver, and immune system, as well as improving mental function, cholesterol levels, digestion, hypertension, tooth care, diabetic conditions, and even aging. It sounds like a miracle elixir, but many studies have had inconsistent results, and the chemical mechanisms that go on in our bodies are not well understood.

To fight cholesterol and blood clots, it's thought that something in tea (it's not yet clear exactly what, but catechins are suspected) works to prevent cholesterol from accumulating in the bloodstream, possibly preventing strokes and heart attacks. Theanine, specifically, is suspected to boost cardiovascular activity and immunity. A number of vitamins are also present in tea, particularly vitamin C, which may also help with boosting the immune system. But more detailed studies need to be conducted to confirm these claims and theories, and to determine the exact chemical mechanisms by which tea operates in our bodies to bring this all about.

While chemicals in tea have been extensively studied in laboratories and these experiments have suggested that these chemicals have various beneficial properties, understanding how the chemicals actually function in something as complex as a human body is a very different matter. In addition to performing the amazingly difficult task of parsing apart how one small aspect of a person's diet affects their health, researchers also have to take into consideration how, for example, differences in lifestyles further complicates the matter. It's a daunting endeavor. But it's important to keep funding research even if it may not seem very relevant at the time; it's hard to predict where the cure for a currently incurable disease, such as cancer, may lie in wait for discovery.

Further Reading

The following resources may be of interest for those wanting to learn more about tea and its biochemistry:

- Mary Lou Heiss and Robert J. Heiss' book *The Story of Tea*
- Kit Chow and Ione Kramer's book *All the Tea in China*
- Bennett Alan Weinberg and Bonnie K. Bealer's book *The World of Caffeine*
- Yukihiko Hara's book *Green Tea: Health Benefits and Applications*
- Chi-Tang Ho, Jen-Kun Lin, and Fereidoon Shahidi's book *Tea and Tea Products: Chemistry and Health-Promoting Properties*
- Joe Schwarcz's books *Dr. Joe & What You Didn't*, *An Apple a Day*, and *Science, Sense, and Nonsense*
- The U.S. Food and Drug Administration (FDA)'s news release "FDA Issues Information for Consumers about Claims for Green Tea and Certain Cancers" at http://www.fda.gov/NewsEvents/Newsroom/PressAnnouncements/2005/ucm108452.htm
- The University of Maryland's Medical Center's webpage "Green Tea" by Steven D. Ehrlich at http://www.umm.edu/altmed/articles/green-tea-000255.htm

Special thanks to Marlon Molinare of Chan Teas for feedback on this essay.

Save the Seeds

As crops become increasingly threatened, seed banks become increasingly important

People have been saving seeds for some 10,000 years, though the practice has become threatened due to recent changes in agricultural practices. To ensure the future of our crops, it's now all the more important to bank our seeds.

The development of agriculture stems from when people discovered the great power of saving seeds. It's been for some 10,000 years that people have been carefully collecting and saving seeds from a crop one year so that they could be used to plant another crop the following year. But agricultural practices have dramatically changed over the last century, and this foundational act of seed saving has nearly become a lost art over the past couple decades. Seeds are increasingly not being saved from year to year, as traditional, local crops are being replaced by identical, new crops in fields all over the world. This development threatens tens of thousands of plant species with extinction and is causing many to once again recognize the importance of saving seeds, not just from local fields, but from all around the world.

THE HISTORY OF SAVING SEEDS

Saving seeds is an essential component of agriculture itself. Originally, it's thought that people collected seeds from edible wild plants and learned how to grow these plants in a more controlled fashion so they didn't have to solely rely on gathering food from the wild plants. By selecting seeds from the most desirable plants (those with the largest fruits or other edible parts, or that simply grew best in the local area), people over time made domesticated crops that were a genetic improvement upon their wild relatives, or an "improvement" at least as far as the people who wanted to eat them were concerned. Over thousands of years, this selection process led to the amazing array of crops we have today, with people historically consuming over seven thousand plant species in their diets.

But over the last couple decades farmers have increasingly broken from this time-honored practice of saving their seeds, instead buying them commercially. Usually these commercially available seeds produce plants with particularly alluring traits, such as bearing more fruit or being more robust. However, in many cases the plants from these commercial seeds do not breed "true" from one generation to the next: Hybrid seeds produce plants that are genetically uniform the first generation, but their "offspring" don't retain their vigor or desirable traits. Consequently, farmers become dependent on buying new seeds each year. As if the genetic deterioration weren't enough incentive, in most cases it's also illegal for farmers to save the seeds of plants grown from commercial seeds, or even for their neighbors, who didn't purchase the seeds, to accidentally have some seeds land on their property and sprout.

While I was a graduated student at the University of California, Santa Barbara, in 2011 this matter received attention in the *Santa Barbara Independent*, as organic farmers from Santa Barbara, along with a total of 60 others from around the country, filed a lawsuit against Monsanto, the global leader in supplying genetically modified (GMO) crop seeds. For more on this lawsuit, see http://www.independent.com/news/2011/apr/15/organic-farmers-sue-gmo-juggernaut-monsanto/. Some of Monsanto's most famous seeds are "Roundup ready," which produce plants that have been made resistant to the broad-spectrum herbicide Roundup, so that farmers can simply spray with this herbicide to eliminate weeds while the crops are spared. It's illegal for farmers who did not purchase the GMO seeds to grow them in their fields, but the GMO crops show up in fields whose owner's claim they did not use Monsanto seeds. In the U.S. alone, despite their bounty and utility, problems with Monsanto crops have affected some tens of thousands of farmers. The widespread cultivation of GMO crops is just one more reason why people are focusing more efforts on creating seed banks.

MORE REASONS TO BANK SEEDS

Long before the development of widely available commercial crop seeds, the Russian biologist Nikolai Vavilov (1887-1943) first thought up, and created, a large seed storage bank. Vavilov gathered seeds globally, traveling for decades over five continents and collecting a wider variety of seed species than anyone else had ever amassed. His research was so important to his group that, during the Siege of Leningrad in World War II, an assistant of his died of starvation rather than eat the edible seeds in their seed storage facility in Leningrad.

Why is storing seeds so important? A major reason is to prevent species and strains from becoming extinct. As discussed above, more and more framing operations have switched from their traditional, local crop varieties to using relatively few varieties that are commercially available. Consequently, farmers are increasingly no longer saving and maintaining the seeds of the old varieties. Additionally, societal upheaval from wars or natural disasters can have a similar effect. It's hard to know exactly how many crop varieties have already been lost, but some believe that number may be in the thousands. Diversity on the species level is most certainly being lost as well: Humans once had around 7,000 plant species in their diets, while today we grow less than 150, and only 12 make up the majority of what we actually eat. And once these species and varieties are gone, they're gone forever.

One might think that the loss of some crop varieties isn't such a terrible thing. Surely the new varieties we've developed, and focused on using today, are an improvement upon the old ones, so what's to miss? These new varieties may seem "better," in that they may produce more fruit, or resist certain insects or diseases, or tolerate harsher climates, but for an organism to truly stand the test of time it needs diversity in its gene pool.

It's hard to predict what challenges an organism will face in the future that it hasn't encountered yet. For example, when a strain of deadly flu initially contacts a population of people, some people will be more susceptible to infection than others. We have this varying range of vulnerability to infection in large part because our genetic pool is diverse. But if we didn't have this diversity, and everyone was susceptible to the new viral strain, it could wipe out the human race.

It's the same with crops. By focusing on growing one variety, or even just a few varieties, of a crop, we run a great risk of having an entire crop succumb to devastation, such as we saw with the near-extinction of the Gros Michel bananas due to the Panama disease (a fungal wilt), which is now wreaking havoc on the Cavendish bananas, which are the sole type of banana that most of us eat. For more discussion on this, see http://www.the-scientist.com/?articles.view/articleNo/30953/title/The-Beginning-of-the-End-for-Bananas-/. It's essential to maintain a large amount of genetic diversity to adapt to all kinds of unanticipated changes in the future.

Researchers working at seed banks not only want to maintain this genetic diversity in crops, but they also use it to create new varieties that can handle current and unforeseen agricultural challenges. This is actually the primary reason Vavilov collected all those seeds; he was determined to breed different varieties together to improve upon existing crops. The main difference between these efforts and the ones generating commercial hybrid seeds is that the former aims to create a wide range of "true breeding" varieties that are ready to deal with all kinds of environmental challenges. For example, some varieties may have the ability to withstand climate change, such as droughts or heat, or changes in soil conditions, like increased salinity. Crops can even be bred that require less land or water, improving the agricultural footprint needed to feed people and saving forests from deforestation.

HOW THE BANKS WORK

Today, there are about 1,400 seed banks in over 100 countries globally. It's estimated that there are some six million samples of seeds in all of these banks, but due to overlap in the seed types collected, only around one to two million are distinct.

How do these "banks" store all this reproductive material? The seeds are collected in fields by researchers, or sent in through correspondence. Some 90% of plant seeds store well in very cold, dry conditions, specifically from about 14 to -4 degrees Fahrenheit. Storing each sample in its own sealed, airtight container this way is a practical and economical way to keep viable reproductive material from the vast majority of plants globally.

But while this seed storage method works well for most plants, it doesn't work for all. Some seeds, known as recalcitrant seeds, die if exposed to cold temperatures (below 50 degrees Fahrenheit) or if they're dried out. Many tropical and subtropical plants have recalcitrant seeds; one such plant near and dear to Southern Californians is the avocado. Because it is so difficult to store recalcitrant seeds long-term, to keep these genetics active in the seed banks other reproductive plant parts or even complete plants must be maintained. But like most things in biology, seed storage isn't black or white; some seeds tolerate a range of temperature and desiccation storage conditions in between these two seed groups.

In ideal storage conditions, some crop seeds, such as peas, may last 20 to 30 years, while others, including many types of grains, can still be viable after hundreds of years of storage. Outside of the carefully controlled conditions in storage facilities, researchers discovered seeds from a lupine plant (*Lupinus arcticus*) in Canada beneath permafrost that were thought to be 10,000 years old that still, amazingly, produced healthy plants, testifying to the ability of seeds to withstand the tests of time. It's thought that seeds' decreased viability over time is not due to using up food reserves, but is due to enzymes ceasing to function that are necessary for the seed to use these reserves, along with DNA repair mechanisms breaking down. To ensure that viable seeds are kept in the seed banks, researchers continually take a few seeds from stored samples, grow them up, collect the new seeds, and put these back into storage.

THE SVALBARD GLOBAL SEED VAULT

As an extra backup measure for all seed banks, the Norwegian Svalbard Global Seed (SGS) Vault was built. Opened in 2008, SGS has the capacity to store duplicates of all the unique seed samples from seed banks around the world; it can hold over four million different seed samples, and all depositors retain rights over, and access to, their seeds. This construction is especially important because up to half of the seed banks in developing countries are threatened by political instability or natural environmental problems, and it is difficult for many third world countries, which have the most plant diversity, to afford seed bank facilities. Located on the Norwegian archipelago of Svalbard, just 600 miles from the North Pole, SGS is under a permafrost mountain, so that even if it lost power the seeds would remain below 26 degrees Fahrenheit. And walls over three feet thick with steel-reinforced concrete protect it from other natural disasters, ensuring the continuation of our 10,000-year-old practice of seed saving. For more information on SGS, see

http://www.regjeringen.no/en/dep/lmd/campain/svalbard-global-seed-vault.html.

HOW TO HELP

While the creation of the Svalbard Global Seed Vault and the 1,400 other seed banks in the world are tremendous accomplishments, it's important to keep these efforts going, especially in the face of global climate change and societal upheaval. The biggest challenges seed banks in general face are a lack of funding and resources. With the amazing power of the Internet, it usually doesn't take too much searching to find relatively local seed banks and learn about what they're doing.

For example, for information on a Californian seed bank associated with the California Rare Fruit Growers organization, see http://www.crfg.org/seedbank.html. And if you're a gardener, you can help maintain rarer plant varieties by growing them in your garden and saving your own seeds. To look at catalogs of heirloom seeds online, see http://rareseeds.com. To learn more about saving seeds, see http://www.seedsave.org/issi/issi_904.html. Local seed exchanges are fairly common; usually anyone can join in and swap some seeds, continuing an agricultural tradition that's been around for thousands of years.

Further Reading

The following resources may be of interest for those wanting to learn more about saving seeds:

- Bonwoo Koo, Philip Pardey, Brian Wright, et al.'s book *Saving Seeds*
- N. Rao, Jean Hanson, M. Dullo, Kakoli Ghosh, David Nowell, and Michael Larinde's book *Manual of Seed Handling in Genebanks*
- Mary Leck, V. Parker, and Robert Simpson's book *Ecology of Soil Seed Banks*
- Genetic Resources Action International (GRAIN)'s website at http://www.grain.org
- The International Seed Federation's website at http://www.worldseed.org
- The International Seed Saving Institute's website at http://www.seedsave.org
- The Svalbard Global Seed Vault's website at http://www.regjeringen.no/en/dep/lmd/campain/svalbard-global-seed-vault.html

Image Credits

Except where noted, all images were taken from Wikimedia Commons and redistributed freely as they are in the public domain. Teisha J. Rowland would like to apologize if any errors have been made in the image credits of this book and corrections can be made in future editions.

Covers and pg. 37, Blood cells (SEM blood cells.jpg), Bruce Wetzel and Harry Schaefer, National Cancer Institute (http://commons.wikimedia.org/wiki/File:SEM_blood_cells.jpg).

Covers and pg. 56, Neuron (Neuron with oligodendrocyte and myelin sheath.svg), LadyofHats and Andrew c (http://commons.wikimedia.org/wiki/File:Neuron_with_oligodendrocyt e_and_myelin_sheath.svg).†

Covers and pg. 121, Sperm and egg (Sperm-egg.jpg), http://www.pdimages.com (http://commons.wikimedia.org/wiki/File:Sperm-egg.jpg).

Pg. 3 and pg. 15, Human embryonic stem cell colony (Humanstemcell.JPG), Ryddragyn (http://commons.wikimedia.org/wiki/File:Humanstemcell.JPG).

Pg. 11, Stem cell self-renewal and differentiation diagram, Teisha J. Rowland*

Pg. 17, Blastocyst (Blastozyste.svg), Lennert B (http://commons.wikimedia.org/wiki/File:Blastozyste.svg).†

Pg. 23, Human eye (My eye.jpg), Gtanner (http://commons.wikimedia.org/wiki/File:My_eye.jpg).

Pg. 29, Brain (MRI brain.jpg), Fastfission (http://commons.wikimedia.org/wiki/File:MRI_brain.jpg).

Pg. 32 and pg. 33, Hematopoiesis (Kit expression in hematopoietic cells.JPG), Ludachris55 (http://commons.wikimedia.org/wiki/File:Kit_expression_in_hematopoi etic_cells.JPG).†

Pg. 43, Tooth diagram (Cavities evolution 1 of 5 ArtLibre jnl.png), Jean-no (http://commons.wikimedia.org/wiki/File:Cavities_evolution_1_of_5_A rtLibre_jnl.png).†¹

Pg. 49, Uterus and uterine tubes (Illu cervix.jpg), Speck-Made (http://commons.wikimedia.org/wiki/File:Illu_cervix.jpg).†

Pg. 61, Dolly the sheep (Dollyscotland (crop).jpg), TimVickers (Llull) (http://commons.wikimedia.org/wiki/File:Dollyscotland_(crop).jpg).†

Pg. 63, Cloning of Dolly the sheep (Dolly clone.svg), Squidonius (http://en.wikipedia.org/wiki/File:Dolly_clone.svg).

Pg. 69, Hair follicle diagram, Teisha J. Rowland*

Pg. 77, Lungs (Lungs.gif), Mikael Häggström (http://commons.wikimedia.org/wiki/File:Lungs.gif).

Pg. 83, Salamanders (Ambystoma mexicanum at Vancouver Aquarium.jpg), ZeWrestler (http://commons.wikimedia.org/wiki/File:Ambystoma_mexicanum_at_Vancouver_Aquarium.jpg).[2]

Pg. 87, Jellyfish (Turritopsisnutricula.jpg), VitorLinhares (http://commons.wikimedia.org/wiki/File:Turritopsisnutricula.jpg).

Pg. 95, Old microscope (Hooke-microscope.png), Robert Hooke (http://commons.wikimedia.org/wiki/File:Hooke-microscope.png).

Pg. 97, Section of cork (Cork Micrographia Hooke.png), Robert Hooke (http://commons.wikimedia.org/wiki/File:Cork_Micrographia_Hooke.png).

Pg. 101, Animal cell diagram (Animal cell structure de.svg), LadyofHats (Mariana ruiz) (http://commons.wikimedia.org/wiki/File:Animal_cell_structure_de.svg).†

Pg. 107, Fruitfly (Drosophila-melanogaster-Nauener-Stadtwald-03-VII-2007-10.jpg), Botaurus (http://commons.wikimedia.org/wiki/File:Drosophila-melanogaster-Nauener-Stadtwald-03-VII-2007-10.jpg).

Pg. 113, Tetrahymena (Tetrahymena thermophila.png), Robinson R (http://commons.wikimedia.org/wiki/File:Tetrahymena_thermophila.png).[3]

Pg. 115, Early embryo (Embryo, 8 cells.jpg), Ekem (http://commons.wikimedia.org/wiki/File:Embryo,_8_cells.jpg).

Pg. 129, Corn (Corn tassels.png), Huw Williams (Huwambeing) (http://commons.wikimedia.org/wiki/File:Corn_tassels.png).

Pg. 137, E. coli (EscherichiaColi NIAID.jpg), NIAID (http://commons.wikimedia.org/wiki/File:EscherichiaColi_NIAID.jpg).†

Pg. 145, Cow (Blaarkop.JPG), Tijbbe (http://commons.wikimedia.org/wiki/File:Blaarkop.JPG).

Pg. 155, Person administering a vaccine (US Army 51521 U.S. Army sponsors first HIV vaccine trial to show some effectiveness in preventing HIV.jpg), United States Army (http://commons.wikimedia.org/wiki/File:US_Army_51521_U.S._Army_sponsors_first_HIV_vaccine_trial_to_show_some_effectiveness_in_preventing_HIV.jpg).

Pg. 161, Lung cancer (X-ray(Chest)Cancer.jpg), National Cancer Institute (http://commons.wikimedia.org/wiki/File:X-ray%28Chest%29Cancer.jpg).†

Pg. 169, Blastocyst picture (Blastocyst, day 5.JPG), Ekem (http://commons.wikimedia.org/wiki/File:Blastocyst,_day_5.JPG).

Pg. 175, Medulla blastoma (CompT1.jpg), Reytan (http://commons.wikimedia.org/wiki/File:CompT1.jpg).†

Pg. 181, Mice (Lightmatter lab mice.jpg), Aaron Logan (http://commons.wikimedia.org/wiki/File:Lightmatter_lab_mice.jpg).[4]

Pg. 187, Ducks (Taiwanese duck farm.jpg), MiNe (sfmine79) from Taipei, Taiwan (http://commons.wikimedia.org/wiki/File:Taiwanese_duck_farm.jpg).[5]

Pg. 197, Glass of port (Port wine.jpg), Jon Sullivan (www.pdphoto.org) (http://commons.wikimedia.org/wiki/File:Port_wine.jpg).

Pg. 205, Tea plant (Camellia sinensis Japan.JPG), Qwert1234 (http://commons.wikimedia.org/wiki/File:Camellia_sinensis_Japan.JPG).

Pg. 213, Pouring tea (Teapause.jpg), Kristina Walter (http://commons.wikimedia.org/wiki/File:Teepause.jpg).

Pg. 221, Seeds (Rice grains (IRRI).jpg), IRRI Images (http://www.flickr.com/people/86712369@N00) (http://commons.wikimedia.org/wiki/File:Rice_grains_%28IRRI%29.jpg).[5]

* Image copyright Teisha J. Rowland.

† Image was modified by Teisha J. Rowland.

[1] Image was redistributed freely as it is under the Free Art License.

[2] Image was redistributed freely with author attribution as it is under the Creative Commons Attribution 3.0 Unported License.

[3] Image was redistributed freely with author attribution as it is under the Creative Commons Attribution 2.5 Generic License.

[4] Image was redistributed freely with author attribution as it is under the Creative Commons Attribution 1.0 Generic License.

[5] Image was redistributed freely with author attribution as it is under the Creative Commons Attribution 2.0 Generic License.

Index

A

adipose tissue (fat tissue), 5, 13, 44, 45, 51, 139, 217

alcohol, 140, 157, 197–203

  agave cactus, 202

  brandy, whiskey, gin, and rum, 201

  cider, 201

  history, 197–203

  mead and kvass, 201

  sake, 202

  wine and beer, 198–200

B

blastocyst, 5, 6, 17, 18, 62–67

bioengineering, 7, 77–81

  Lung bronchus, 7, 77–81

  Castillo, Claudia, 7, 77–79

  Macchiarini, Paolo, 78–81

bone marrow, 4, 5, 7, 13, 38, 44, 49, 79, 80, 124

C

cancer, 3, 4, 8, 11–14, 35, 37–41, 58, 90, 110, 113, 124, 155–194, 217, 219

  cancer stem cells, 8, 11–14

  cancer vaccines, 8, 14, 155–174, 189, 191, 193

  cell markers, 13, 14

  childhood cancers, 175–179

  mutations, 12, 175–179, 183

  oncogenes, 12

  telomerase, 58, 181–186

  teratomas, 19

  tumor suppressor genes, 12, 178

  leukemia, 4, 12, 33, 37–41, 146

  lymphoma, 4, 33, 37–41

  viruses, 156–159, 162, 164, 166, 189, 191, 193, 194

cell markers, 13, 14

cells, discovery of, 95–105

cell structure (animal cells), 95–105

  cell membrane, 102, 103

  chromosomes, 102, 108, 109, 123, 181–183

  endoplasmic reticulum, 101, 104, 105

  Golgi apparatus, 101, 103, 105

  mitochrondria, 65, 101, 103–105

  nucleus, 25, 26, 62–66, 98, 99, 101–103, 105, 108

T

www.ingramcontent.com/pod-product-compliance
Lightning Source LLC
Chambersburg PA
CBHW031833170526
45157CB00001B/289